Ecology, Ethics, and the Future of Humanity

PALGRAVE STUDIES IN THE FUTURE OF HUMANITY AND ITS SUCCESSORS

Series Editors: Steve Fuller and Calvin Mercer

Humanity is at a crossroads in its history, precariously poised between mastery and extinction. The fast-developing array of human enhancement therapies and technologies (e.g., genetic engineering, information technology, regenerative medicine, robotics, and nanotechnology) are increasingly impacting our lives and our future. The most ardent advocates believe that some of these developments could permit humans to take control of their own evolution and alter human nature and the human condition in fundamental ways, perhaps to an extent that we arrive at the "posthuman" or the transhuman, the "successor" of humanity.

This series brings together research from a variety of fields to consider the economic, ethical, legal, political, psychological, religious, social, and other implications of cutting-edge science and technology. The series as a whole does not advocate any particular position on these matters. Rather, it provides a forum for experts to wrestle with the far-reaching implications of the enhancement technologies of our day.

Transhumanism and the Body: The World Religions Speak
Edited by Calvin Mercer and Derek F. Maher

Ecology, Ethics, and the Future of Humanity
Adam Riggio

Ecology, Ethics, and the Future of Humanity

Adam Riggio

palgrave
macmillan

ECOLOGY, ETHICS, AND THE FUTURE OF HUMANITY

First published in 2015 by
PALGRAVE MACMILLAN®
in the United States—a division of St. Martin's Press LLC,
175 Fifth Avenue, New York, NY 10010.

Where this book is distributed in the UK, Europe and the rest of the world, this is by Palgrave Macmillan, a division of Macmillan Publishers Limited, registered in England, company number 785998, of Houndmills, Basingstoke, Hampshire RG21 6XS.

Palgrave Macmillan is the global academic imprint of the above companies and has companies and representatives throughout the world.

ISBN: 978–1–137–54075–1

Library of Congress Cataloging-in-Publication Data

Riggio, Adam, 1983–
Ecology, ethics, and the future of humanity / Adam Riggio.
pages cm —(Palgrave studies in the future of humanity and its successors)
Includes bibliographical references and index.
ISBN 978–1–137–54075–1—

1. Human ecology—Moral and ethical aspects. 2. Environmental degradation—Moral and ethical aspects. 3. Climatic changes—Moral and ethical aspects. 4. Environmental ethics. I. Title.

GF21.R54 2015
179′.1—dc23 2015007244

A catalogue record of the book is available from the British Library.

Design by Newgen Knowledge Works (P) Ltd., Chennai, India.

First edition: August 2015

10 9 8 7 6 5 4 3 2 1

Dedicated to
James Bradley
and
Arnold Bennett
Two old friends and teachers.
I so looked forward to sharing this with both of you.

Contents

Acknowledgments

This book began as a doctoral dissertation, and is more ambitious than most of them. So I am primarily thankful to my supervisor Barry Allen at McMaster University, for accepting and supporting the scope of my plans. The other professors who served on my committee, Diane Enns, Nancy Doubleday, and Brigitte Sassen, have also made important contributions to this project in the early stages of its development.

The support of Steve Fuller of the University of Warwick was essential to getting *Ecology, Ethics, and the Future of Humanity* published in the first place. His organization and active role in the Social Epistemology Review and Reply Collective (SERRC) has fostered thinkers and scholars around the world. SERRC has been my institutional home in humanities research and writing since leaving McMaster, so I also owe thanks to its organizers, James Collier and Sue Dieleman, and my regular collaborators over the years, especially Emilie Whitaker, David Butz Pedersen, Francis Remedios, Elisabeth Simbürger, Thomas Basbøll, and Todd Suomela. Palgrave Macmillan staff have given invaluable support to the project as I prepared the book in its current form.

Most of the research for *Ecology, Ethics, and the Future of Humanity* took place at McMaster University, whose Philosophy department was my base for five years. So thanks are due to that department's other professors while I worked there—Richard Arthur, Nicolas Griffin, David Hitchcock, Violetta Igneski, Mark Johnstone, Stefan Sciaraffa, Wil Waluchow, and Elisabeth Gedge, the department chair at the time. I also thank the department administrators, Daphne Kilgour and Kim Squissato.

Other research institutions have supported my work over the years and offered me invaluable education and experience. I began my philosophical studies at Memorial University, where the philosophy department's classes, seminars, and discussion groups introduced me to so much of what philosophical practice can achieve. My teachers there deserve my most sincere gratitude. They include my master's supervisor Arthur Sullivan, and the department's professors Craig Cramm, Jennifer Dyer, Jay Foster, Sheldon Hanlon, Peter Harris, the late Darren Hynes, Jennifer Flynn, Richard Matthews, Natalie Oman, Walter Okshevsky, Seamus O'Neill, Suma Rajiva, John Scott, Evan Simpson, Toni Stafford, David Thompson, and Bernie Wills. Also noteworthy are my fellow students from that program who became my good friends—Kyla Bruff, Joseph Carew, Ray Critch, Elizabeth Doyle, Harold Duggan, Billy Kennedy, Aaron O'Brien, Lynn Panting, Gil Shalev, and Danielle Sullivan.

From 2009 to 2014, I regularly attended the annual conference of the Canadian Philosophical Association, which Louise Morel of Saint Mary's University has regularly coordinated. I attended the Summer Institute in American Philosophy in 2011, where I learned from Joe Margolis of Temple University, Colin Koopman of University of Oregon, Brad Stone of Loyola Marymount University, and Eric Weber of University of Mississippi. I also attended multidisciplinary conferences at the University of Edinburgh, University of St. Gallen, University of Cuenca, and University of Toronto organized by Common Ground Conferences. So I thank Homer Staveley, Audrey LeGrande, and Emily Kasak at Common Ground.

As well, I would like to thank Manuel DeLanda for his personal correspondence and advice during the writing of this manuscript. John McLevey and Allyson Stokes of the University of Waterloo have been fellow researchers and close friends for many years.

Many friends and fellow students at McMaster have contributed to my thoughts as I put together this book. They include Justine Ajandi, Chris Alaimo, Russell Anderson, Ria Angelo, Scott Balasak, Dorian Batycka, Mike Bennett, Patrick Bondy, André Bovet, Jessica Bowen, Shannon Buckley, Geoff Callaghan, Dave Campbell, Matt Coombe, Darren Cyr, Elizabeth Da Ponte, Dave Devlaeminck, Jorge

Fabra, Andrew Garrod-Martin, Matt Grellette, Nicole Guerreiro, Kayte Hachey, Qiang Hao, Michael Hemmingsen, Noel Iverson, Veronica Jefferies, Laura Kennedy, Karl Laderoute, Lucy Langston, Jared Lecker, Harrison Lee, Jeremy Livingston, Brynna Loppe, Ed Lovo, Arthur McLuhan, Joshua Mildenberger, Emily Milne, Jessica Murphy, Anca Nichifor, Dustin Olson, Igor Osowski, Don Oxtoby, Otto Phillips, Kait Pinder, Tano Posteraro, Jennifer Primmer, Borna Radnik, David Rondel, Fábio Schecaira, James Sikkema, Shen Storm, Alanda Theriault, Benjamin Wald, Scott Wisdom, Yussif Yakubu, Joanna Zaslow, Joshua Zaslow, and Haixia Zhong.

And Gillian, for her uplift and love.

Introduction: A New Human Emerges from Ecological Disaster

"Interesting philosophy is rarely an examination of the pros and cons of a thesis. Usually it is, implicitly or explicitly, a contest between an entrenched vocabulary which has become a nuisance and a half-formed new vocabulary which promises great things" (Rorty, 1989, p. 9). Although Richard Rorty here speaks explicitly about ways of speaking, he clearly describes philosophy's simplest and most profound purpose: developing new ways to be human. *Ecology, Ethics, and the Future of Humanity* is part of a series on posthumanist thought, and it is also a critique of the prevailing trends of writing that identifies itself as posthumanist or transhumanist. My critique is not an explicit argument with any writers and thinkers in the field, simply a statement. A new paradigm of humanity will only be effective if it resolves a practical problem. Transhumanism for the sake of transformation alone is little more than an empty philosophical masturbation. This book discusses a new conception of humanity that is emerging from the problem of our planetary ecological crisis.

We will explore an ecological conception of what humanity is and can do. Imagining a new humanity with this inspiration will completely transform our popular conceptions of self-interest, the human sense of home, and our material self-conception. The arguments of this book develop a set of principles in a variety of contexts that together form a philosophical framework for an ecological conception of humanity. One principle is that all bodies are singular. Contingent processes assemble a body with a singular identity that is more specific than any of the general concepts of philosophical language one can use in an account of that body. Selves are one kind of these contingent, singular bodies. Jean-Paul Sartre, in his novel *La Nausée*, describes this contingency and the inability of general propositions to give a complete account of a singularity, as the excess of anything that exists over any definition we can give of it (Sartre, 1964, pp. 126–135). This contingency disturbs the narrator Roquentin because the necessity of a comprehensively true explanation or account acts as an anchor of certainty that grounds his thinking and life. There is no certainty in an entirely contingent world where the work of understanding can never be complete. The continuing generation of novelty inevitably overcomes the apparent completion of human knowledge. We therefore live in a state of constant vigilance for the generation of new problems that our familiar concepts and habits of thought are unable to handle. We could fear these new problems for how they unsettle the presumptions that once comforted us. But we can develop the strength to accept them as an invitation to explore the complexity of a diverse universe.

That we should value and protect the world's diversity is a common principle in environmentalist moral philosophy. So any new vision of humanity that emerges from responding to our ecological crisis should consider this value fundamental to its own moral sensibility. The concept should be more than a simple exhortation to value diversity, because it would otherwise be a mere dogma. Encouraging diversity goes beyond simply preserving what already exists, because while preservation is important, to preserve alone stagnates an ecosystem. We must therefore act so that the world becomes more diverse. Because the dynamisms of ecosystem are the product of interactions and the growth of diversity depends on these interactions, interaction alone fundamentally integrates everything that exists, including humans.

This means that the ecological human must reject another contemporary environmentalist dogma, which has the deepest roots of all. The common belief that humanity, our civilization, and our technology are essentially and incurably separate from nature is not only false, but also dangerous. Such a worldview prevents us from understanding how deep and complicated our links, integration, and interdependence are with the ecosystems in which we grew up. A conception of the human relationship with nature as an absolute dichotomy, two poles of being that are forever apart and opposed, can result only in unintentional mass suicide or a depressed fatalism that Man and Earth can never be reconciled and that one will eventually destroy the other. Humans have always been technological: our peculiarly plastic neurology integrates us with our environment and enables the intense creativity of our tool use. Nature is no Eden from which humanity has fallen because of our technological innovations. Human ingenuity has created the enormous industry whose pollution, ecosystemic destruction, and transformation of the planet's climate may likely destroy humanity itself. We embody the most theatrical maladaptation in Earth's history. Yet for all the terrifying beauty of our most fantastic industrial creations, our danger is neither profound nor inevitable. We are drowning in our own shit.

The essence of the danger humanity faces from its global technological stumble is no less ordinary than our mere death as a species, and we have caused enough of that. But the remedy for this behavior must profoundly transform our fundamental attitudes about what humanity is and what makes a human great. We must leave the massive achievements that dominate a complex land to overlay a unified human order, whether the phallic joke of a skyscraper or the life-draining blandness of monoculture industrial farming, in the past. Humanity's survival depends on crafting a new sense of humility in understanding our ecological roles and functions. I do not intend to return to any previous humility, where nature was regarded with the fear appropriate to an unknown sublime. We can no longer fear nature as we feared the wrath of gods. That attitude was just as alienated from our ecological integration as humanity's current self-image as technological supermen; we merely cowered instead of conquered. Humanity must educate itself as a people about ecological processes and how our existence can encourage the diversification through which ecosystems thrive. Knowledge must humble us.

At the same time, we must be careful not to force ecological moralities on people through the coercion of government. Most of our environmental laws depend on identifying and punishing people and companies who commit ecologically destructive acts. Such laws are effective in the short term, but they also breed resistance because those who suffer from their punishments do not understand what they have done wrong. If someone does not understand what harm there is in cutting down a forest to bolster immediate economic growth, then a state law will not teach them. The coercion of the state to enforce an environmentalist society will eventually generate popular resistance to environmentalism itself because humanity chafes under chains. Effective environmentalism requires each individual person to join the movement herself, changing her morality and her entire sense of self. This is not a path of coercion, but of inspiration. Environmental moral norms flow naturally from a conception of yourself as an ecological creature whose intimate connection with your surroundings is just as essential to who and what you are as your thoughts themselves.

I hope this book provides one such means of inspiration. Each of its seven chapters discusses a separate topic, but each successive chapter revisits the same core ideas. They are creative repetitions, as working through a concept once changes how you understand and use it when it appears again in a different problem. Philosophy typically has very strict disciplinary boundaries, which this book generally ignores. *Ontology* inquires exclusively into what is, and the necessary or contingent properties and possibilities of existence. Problems of *epistemology* inquire into the methods and accuracy of understanding experience and information. *Ethics* inquires how bodies capable of self-conscious control can articulate their lives, how one's understanding of self generates an individual and social identity. Ethical thinking has an ontological component, but a tighter focus, discussing what a person is, the nature of a person's identity, how it physically comes to be, or how you understand your personality.

This departs from how we typically think of ethics in disciplinary philosophy and in everyday life, where the word is synonymous with morality. But a philosophy that aims to transform the world through inspiring people to live and think differently must distinguish thought about self-identity and subjectivity from more general questions about existence, and distinguish thought about personal integrity and authenticity from questions of what is right and wrong. Spinoza carried out ethics this way throughout the last two chapters of his *Ethics*. Henri Bergson defined the domains of morality and ethics explicitly in his *Two Sources of Morality and Religion*: normative morality facilitates social cohesion, and ethical inspiration facilitates social progress. Gilles Deleuze and Félix Guattari's magnum opus *A Thousand Plateaus* is a book of experiments in profoundly transforming subjectivities. New moral ideas fail unless people dedicate themselves to reshaping society according to their guidelines from the individual to the civilizational level. A *morality* is a system of rules or principles of actions and responsibilities, which means that political philosophy is a subcategory of morality. A moral inquiry is exploring whether particular principles to guide behavior and responsibilities can together constitute a single internally consistent system.

Metaphysics is a word that most often refers to ontological concerns, but given how my project maps philosophical domains, I use it in a different sense that keeps its traditional sense of grandeur and scope. One implication of understanding philosophy as a subject with demarcated domains is that no one domain is foundational for the others. The demands of your particular inquiry will ask what a domain's principles are prior to its implications for other domains. Most philosophical inquiries restrict themselves to a single domain, but the most ambitious (and the easiest to mess up) connect principles across two or more domains. These complex philosophical assemblages of concepts are *metaphysical* inquiries. This book's inquiry into what concepts would define our ecological self-conception proceeds largely through the interaction of ontological and ethical concepts, drawing moral and political implications.

Ecology, Ethics, and the Future of Humanity begins with critical examinations of major ideas in environmentalist moral philosophy. Their fundamental theme is that people who conceive of themselves as discrete individuals, separate from their environment and having total control over their lives and personalities, can never adopt environmentalist moral principles. Environmentalism understands everything in the world as integrated at all levels of existence, and that our own lives are interdependent with all other lives and systems in our world. Environmentalism is often seen as incompatible with democracy because it emphasizes mutual connections and our dependence on sometimes fragile relationships over individual autonomy and liberty. A fundamental theme of environmentalist moral principles is that we cannot do whatever we want, and that we must care equally or more about how our actions systematically cause suffering than their direct effects, which is where liberal moral philosophy typically focuses.

The first three chapters of the present book consider three ways to articulate this problem that show what a new vision of humanity must achieve to resolve this fundamental crisis of our civilization: in morality, in environmentalism's conflict with liberalism, and in the paradoxes that arise from the practice of abstract thought that grew from an activist movement. The following four chapters build such a conception of humanity literally from the ground up. Chapter four discusses the nature of relation and interdependence itself, how an ecological philosophy should conceive the assemblage of bodies and fields. It is my most explicit argument that a genuinely ecological philosophy is inevitably a process philosophy. A human is, therefore, a process, a becoming, a continuing dynamic interplay of relative fluxes and stabilities. Chapter five describes the simplest structure of perception itself, the autopoietic body, and works out my own interpretation of this complex field of ideas that best achieves an ecological vision of what humanity can become. Chapter six describes how the interactions of organisms with their environment reshapes land to have a perceptual and meaningful character. Chapter seven describes how the peculiar interaction of humans with our environment is a natural process that constitutes our inherently technological subjectivities. As well, I describe how lasting political and social change only occurs through webs of inspiration among individuals, transforming subjectivities one at a time. This is political activism as an ecological process.

A figure who recurs at different points is the Norwegian philosopher Arne Næss, whose writings in his later life explored, but could not escape, many of the paradoxes that arise when advocating an environmentalist morality in a culture that has not yet learned to conceive of the self as genuinely ecological. Many thinkers have been caught in this puzzle, but Næss is a notably poetic example because he was able to see the problems, but not how to overcome them. That path of overcoming finds its clearest signposts in the works of Deleuze and Guattari, whose ideas also recur throughout this book. Recent interpreters of their writings have come to see the ecological character of their ideas, and I count myself among them. More than interpreting their concepts, however, I use them and the fields that influenced them to build a framework of thought that leaves our current conception of selfhood that incurably alienates us from nature behind. We should not reconcile our common self-conceptions with environmentalist moralities, but instead transform how we understand ourselves, transforming what it means to be human.

Chapter 1

Nature's Intrinsic Value: A Forgotten Philosophy of the Environment

Although the concept has faced relative neglect in recent decades and withered under a well-deserved critique, one major goal of this book is to advocate for the intrinsic value of nature as the best foundational principle in a genuinely ecocentric moral philosophy. This is the principle that nature is valuable according to its own essence and attributes, whether or not a subject recognizes it as valuable (Taylor, 1986, p. 73). This principle has been critiqued as practically unworkable and even conceptually unsound. And those critiques are correct: the problems of what has this value property and what does not, whether value is a property of a body itself or derived from some other property, and the chronic inability of a principle of nature's intrinsic value to provide us any guidance in conflicts between human and nonhuman priorities and projects are all valid and inescapable. This chapter discusses how previous articulations of the principle of nature's intrinsic value are limited, and offers my own remix of the idea that avoids these problems. Instead of nature itself, the object of intrinsic value is natural diversity.

Intrinsic Value as a Means to Make Nature Itself Morally Relevant

One reason why I think the intrinsic value principle first caught on as it did was because of its radical character. Such a definitive break from a tradition of moral philosophy entirely focused on the concerns of humans alone (or in the smaller concessions, of duties to nature as an aspect of human virtue) suits the context of the radical politics of the United States in the 1960s, when environmentalism first became a movement noticeable in mainstream culture. Yet even given the concept's radical nature, academic philosophers who have advocated nature's intrinsic value tend to explain it as conforming to moral beliefs held so widely that they have become culturally intuitive. Even though the tradition's most remarkable works are clear breaks from current and past orthodoxies, it is as if philosophical practice itself exerts a pressure not to rock the boat, to maintain continuity with what has come to be accepted as intuitively true. First, I wish to examine three different paths to establish a firm ground for an environmentalist morality. Of these, only the path that Arne Næss represents avoids this conservative tendency. Næss develops concepts with the potential at least for conditioning a radical break from well-established traditions of moral philosophy and common moral thinking. Environmental moral philosophy can inspire a very different way of thinking about the entire universe than many of us now hold. Taking a concept of nature's intrinsic value seriously is the first step in unlocking that potential.

One approach to ground an environmentalist moral philosophy begins with the widespread intuition that intelligence or rationality is the foundation of moral valuation. This is the foundation of Nicholas Agar's argument for life's intrinsic value in the book of the same name: intentional or unintentional harm to intelligent or sentient creatures constitutes a moral wrong. According to everyday common sense, this intuition applies only to humans, as humans are most often conceived as the only genuinely intelligent species. The entire first half of his book constitutes an argument that radically recasts what intelligence technically is so that this intuition can apply equally well to the aye-aye, unremarkable species of moss, bacteriophages, as well as typically majestic creatures like the sequoias and rhinoceros. With a series of principles drawn from studies of biological functionality and the sensitivity of biological bodies to flows of information, Agar concludes that all perceptual activity is representational, insofar as representation is understanding your surroundings in your own terms. If a creature understands the world representationally, then it is sentient or intelligent. If every living creature is intelligent, then every living creature is intrinsically valuable (Agar, 2001, pp. 92–94). Ecosystems and their stability then become instrumentally valuable as the means to support organisms, all of whom are intelligent to some degree. Agar's goal is to form a conceptual bridge from the everyday common sense morality of folk psychology to the radical territory of ecocentric morality (Agar, 2001, p. 101).

Essentially, Agar relies on scientific principles and discoveries to ground an extension of a popular common sense of moral significance bodies that have not historically been so included. A comparison to animal rights/liberation morality is useful to understand the severe limitations of extension logic in moral philosophical reasoning. Animal liberation is a radical morality in its most superficial aspects, extending humanity's moral rights of humans such as bodily inviolability and the duty to alleviate suffering among animals. But animal rights and liberation moralities leave intact the widely accepted systems of rights, norms, and morals themselves. It merely asks us to expand the class of bodies to which such moral principles and imperatives apply. Peter Singer, the paradigm and progenitor of animal liberation, writes, "The shift from a point of view that is disinterested between individuals within a group, but not between groups, to a point of view that is fully universal, is a tremendous change... It is the direction in which moral thought has been going since ancient times" (Singer, 1981, p. 113). The change is tremendous because it has been a major trend of social development. But because Singer frames the expansion of moral standing as the arc of humanity's moral development over the history of our civilizations, this supposedly radical expansion is a conservative process, one more step on an arc of history that human culture has been following inexorably and inevitably since its beginning.

Singer gives a narrative of human history in which the scope of application for moral principles grows as contact with foreign groups becomes more frequent, and people with different ways of life interact more and more frequently. In a highly isolated, tribally organized society, only members of your own tribe are worthy of moral consideration. Foreigners are outside the range of consideration. As people become more interconnected, they become more accustomed to foreigners. The differences in identity that would have set a person beyond moral consideration cease to matter. We accordingly adjust the range to which moral principles apply: instead of Athenians, they refer to Greeks; instead of French, they refer to Europeans; instead of whites, they refer to humanity. Singer's own moral and political project, outlined in *Animal Liberation*, is a further expansion of who is morally considerable. Instead of humanity, moral principles refer to all sentient creatures. The set of the morally significant expands, but the moral principles and their imperatives are themselves unchanged. In this sense, Agar's and Singer's projects are the same.

The only difference is that Agar uses scientific ideas to redefine sentience as a property of all organisms. Singer keeps himself distant from these ideas because his moral concern is to reduce pain and suffering, or at least to avoid and to prevent harm (Llorente, 2009, pp. 62–63). Singer's moral expansion is therefore limited only to organisms having a particular kind of neuro-physiological architecture such that they can feel pain (Singer, 1981, pp. 123–124). Agar's concept is little better: every individual organism is intrinsically valuable in virtue of its intelligence, and he defines intelligence as conative capacity, its striving to stay alive. Because the individual is the seat of value, valuation of whole species and of ecosystems is derived from the value of the individual. Ecosystemic health is instrumentally valuable in crafting a place for individual organisms to live. He develops a formula to calculate how much the death of one individual matters to the health of all the individuals of the species or local community of that species, making the value of the species a function of the values of various individuals who constitute it (Agar, 2001, pp. 149–150). The expansion of the morally valuable in animal liberation's version of the intrinsic value of nature does not go far enough to include collections of bodies, whether or not those bodies are organic.

The core problem of the expansion approach to environmental moral philosophy is that while the set of bodies to which we apply the imperatives of our moral principles changes and grows, the principles and imperatives themselves do not. It is a fundamentally mistaken path to build an ecocentric moral philosophy. There are many reasons why, but for now, the most important reason is that a genuinely ecocentric morality requires us to dispense with the otherwise common sense imperative not to eat that for which we care. One of the most important relationships for the maintenance of ecosystems is that of predator and prey populations. Ecosystemic health requires preserving relationships of hunter and hunted. Therefore, an ecocentric morality dispenses with moral principles under which all applicable bodies are inviolable.

We can understandably consider this a general flaw in ecocentric moralities, a reason why we should not accept them. It definitely stands against common sense notions of what is morally right. However, common sense is not a guide to clear truths that we can easily intuit. Common sense is better described as a signpost to our most stubbornly entrenched habits of thinking. Only in the contemporary era has humanity faced with any serious intensity problems requiring an ecocentric morality. Humanity has developed moral systems, legal institutions, and political regimes centering on rights and the capacity to advocate for them for humans and human powers. Fitting a body whose capacities are radically different from those of humans into existing human institutions is, for many environmentalist philosophers, the best way to build a moral relationship with such a different organism. The phenomenology of such a body may differ from our own so radically that it is impossibly difficult to imagine how it exists from its own point of view. This is the problem Thomas Nagel identified as "what-it-is-like-ness," the limitation of human thought and language to describe an utterly alien body without anthropomorphizing metaphor (Nagel, 2004, p. 531).

Attempts to adapt our established institutions of legal and moral rights have run into practical and conceptual roadblocks, and this will continue. Ecosystems and nonhuman populations do not fit legal institutions developed with only humans in mind. In the early 1970s, Christopher Stone stood before the United States Supreme Court in defense of the Mineral King Valley outside San Francisco. He argued that the ecosystem constituted by the living and landscape inhabitants of the valley itself held legal rights, which obligated people and their courts to recognize them (Stone, 1972, pp. 450–501). But the following decade of philosophical response to this idea tended to conclude that advocacy for valleys was beyond the capacity of human language and understanding. People would merely pretend that they knew what a valley, a sequoia, a cetacean, a minnow wanted or demanded, and extrapolate likely desires from those bodies' behaviors. But without the possibility of accurate, detailed communication, you deal only with sometimes dubious guesses. Humans cannot understand any interests of bodies well enough to become their advocates in a legal system, unless those bodies can have communicated complex concepts with humans (Nash, 1989, pp. 134–135). While a moral philosophy like Agar's can provide us a template for thinking about how to treat creatures with whom communication is impossible, it cannot guide us in the practical matter of incorporating this morality into our entrenched legal institutions.

I am not skilled enough in the art of the law and legislation to draft a legal regime that would authentically express an ecocentric moral philosophy, nor am I naive enough to think that writing the framework of such a regime in one of the many books on environmental politics and philosophy that are continually published would result in its wholesale adaptation. I bring up the subject to show how deeply entrenched exclusively humanistic moralities are in contemporary society, and that an authentic environmentalist philosophy must deal with the challenge of overcoming common sense. Concepts of intrinsic value have the greatest potential to achieve this break. However, there are two ways to understand intrinsic value: as an object's property that emerges from a human act of valuation, or as a discoverable property of individual bodies in isolation (Robert, 2000, p. 193).

Throughout this work, I discuss bodies, their properties, structure, processes of generation and dissolution, interrelation, and integration. Everything is a body. Any physical assembly, any field of force, any process no matter how quickly it is constituted and dissipates, is a body. One way to understand intrinsic value is typified by the phrase "I value this body intrinsically." I understand its many features, how it operates, its structure, its relations to bodies around it. I understand this body in its singularity, that there is no other body exactly identical to it. Each body is singular and no body cannot be singular. I value this body in its singularity and for its singularity.

But the ubiquity of singularity need not make it a meaningless property. At first hearing, it sounds paradoxical "to utter in the same breath, like Balthasar, 'Everything is ordinary!' and 'Everything is unique!'" (Deleuze, 1993, p. 91). Bodies vary in their degrees of difference from others, but the most important idea for the current discussion is that each body is uniquely different, at least in some small way, and that this uniqueness is itself valuable. This is a concept of intrinsic value that embraces difference. Nowhere in this conception of intrinsic value is valuation grounded in any property common with humanity. A body's singular identity makes it an individual in the strongest possible sense: insofar as no body can be identical to any other, each body differs from every other body.[1] I approach the various interconnected inquiries of the rest of this project with the aim to build a comprehensive argument for why and how singularity can be worthy of valuation. But no matter the reason, a person's moral relationship with a body only comes to be when that person declares the body valuable. If a concept of intrinsic value rests primarily on the recognition of such value, I call it a species of the *valuation understanding* of intrinsic value, which ultimately falls short of the philosophical radicalism that environmentalist thinking needs.

Arne Næss, in contrast, sought to preserve the purest conception of nature's intrinsic value for environmental philosophy. He conceived of value as a natural body's property whose existence delivers a moral imperative that a subject preserve it, which I call the *property understanding* of intrinsic value. He built this technical philosophical conception from the subtleties of the Norwegian word *egenverdi* (Rothenberg, 1989, p. 11). Næss argues that if the value of a body rests on a person's recognition or bestowal of value, then supposedly intrinsic value reduces instead to a ground in human thought and will. This contrasts with how environmentalist philosophers typically conceive of their moral systems as allowing nonhumans to speak for themselves (Cheney, 1998, pp. 265–277). We can discover a body's value property through some kind of observation, usually of a special case (Næss, 1984, p. 268).

I should clarify that the property understanding of intrinsic value does not imply that there exists some property of a body that is intrinsically valuable as an end in itself. For example, here is one way to understand Singer-influenced animal liberation morality. Bodies possess a property, the capacity to suffer, emerging from the appropriate neural architecture, which is intrinsically valuable in moral reasoning. Other thinkers such as Agar and Charles Cockell consider conativity a body's intrinsically valuable property, whether as itself or, in Agar's work, as a redefinition of conventional intelligence. On this ground, Cockell advocates that microorganisms are worthy of moral consideration, and has laid out a framework for such a philosophical system (Cockell, 2005, pp. 375–390). J. Baird Callicott proposes that ecosystems are the proper objects of environmentalist moral consideration, those bodies whose activities aid in the maintenance of an ecosystem's harmonious stability being intrinsically valuable (Callicott, 1989, pp. 88–90).

What Is a Natural Body's Property of Value?

A recurrent fear in environmental moral philosophy is that human reason can no longer be fully trusted in practical and political contexts. If a body's valuation depends at all on human thinking, as in what Paul Taylor calls philosophies of an inherent value in nature, then that act of valuation is open to selfish corruption and caprice. An imperative of environmentalist morality is to avoid valuing nature or some essential aspects of nature entirely instrumentally, as if its good were defined solely through providing a human material benefits or satisfying his desires. Yet so much theory seeks a particular property of natural bodies or nature to be the object of human valuation. Recognizing the intrinsic value of intelligence, conativity, relations engendering harmonious stability, or interdependence turns out to be the inherent valuation of just those properties, but not the intrinsic valuation of the bodies themselves. A concept of nature's intrinsic value demands that humans recognize a value that exists prior to our valuation. Næss remains promising in this regard, if difficult to parse, as his concept of *egenverdi* stands separate from his valuation of a body's capacity to live and blossom.

Ecocentric philosophy aims to reorient moral thinking to make nature, not humanity alone, the central figure of concern (Callicott, 1989, pp. 3–4). It was a philosophical reaction to what was perceived to be a crisis. Humanity had developed technology that did immense environmental harm, using it with impunity and shortsightedness. When philosophers such as Næss, Callicott, Bill Devall, and George Sessions were writing ecocentrism's foundational works in the 1970s and 1980s, they often took technology to be an essentially malevolent exploiter of nonhumans (Devall and Sessions, 1985, pp. 5–6). Nonhuman bodies had, according to these writers, long been valued only instrumentally for how they could benefit humans, as if only humans and human goals were intrinsically valuable. The remedy was to advocate that the value of nonhuman bodies, often referred to en masse as simply "nature," does not emerge from an act of human valuation, but already exists to be discovered. Humans could notice this intrinsic value, or not; if not, we had made a mistake in perceiving the world. If a body's value is generated independently of human thought and action, then it must be a mistake to reduce a body's value to some instrumental role in human needs and desires. A human may believe, thanks to technological society's long inculcation of anthropocentric attitudes, that a body's value depends only how humans can use it. But, according to the property understanding of intrinsic value, that is wrong. It is not only a moral error, but also an epistemic and ontological error: failing to know an important property of a body.

For an ecocentric morality, understanding what the best action is requires attention to perspectives different from your own, and the consequences of your actions in the context of those perspectives. Shortsightedness is therefore a great moral failure, whether a result of willful ignorance, or more frequently merely obliviousness. Yet it is in all cases morally blameworthy, a sign of the significant consequentialism of environmentalist moralities. For example, Terry Tempest Williams describes in *Refuge* a hydroelectric project that caused the Great Salt Lake to flood the Great Bear River Migratory Bird Refuge. The humans who planned and built the hydroelectric project think only of human desires for power generation, believing a summary of these desires to provide a complete account of the situation. They find no rival accounts among nonhumans, oblivious as they are to the activities and needs of the Great Salt Lake ecosystem's multifarious creatures, knowledge of which could have built a more comprehensive account. The developers even ignore other human residents of the area, who are opposed to the hydroelectric development (Browne, 2007, pp. 172–181). Each of these bodies—whether bird, plant, fungus, anti-development human—has its intrinsic value: a property that those responsible for the flooding have not even attempted to uncover. Having presumed that such properties do not exist, they do not bother searching for them.

Living according to an ecocentric morality is extremely difficult. It requires humans to understand not only their own perspectives, but those of all other constituents of an ecosystem. This requires careful observation of all these bodies to build complex concepts based on their needs and histories as well as your own. An account of a situation is a tool to understand that situation, and we strive for our accounts to be as accurate and complete as possible. Dealing with other humans in a situation, my account can be quite accurate, because humans can engage in complex conversations with each other, and synthesize individual accounts into a more comprehensive account. But humans cannot engage in complex dialogue with egrets, fire ants, or mildew, for example. In her analysis of the Great Bear River Refuge's destruction, Williams describes the developers as oblivious to the accounts of the nonhuman inhabitants of the refuge. But the reason for their lack of multispecies knowledge was that they learned about the situation only by looking for accounts that were already linguistic, passing over any other method of investigating the ecosystem. Even this collection was greedily selective, listening only to those humans who already agreed with them.

Situations like these demonstrate a practical difficulty that often goes unnoticed. To understand a creature's perspective on a situation, we must have an account of that perspective. But nonhumans cannot speak. Unable to give an account of their lives or claim some right, we can conclude that nonhumans can have no intrinsic moral rights, although we can give them inherent legal rights in human political systems, and appoint advocates to speak for them, to account for their perspectives (Taylor, 1986, pp. 221–223). However, while humans bestow legal and moral rights on nonhumans, their needs and histories are their own, and the rights we create for them must reflect these faithfully. Unable to communicate with nonhumans, we can never be sure of this. Perhaps you could carefully inspect a creature's actions and infer from that enough information to build an account that it would have no powers to create itself. But even though your account is about the actions of a nonhuman, a human is still creating the account *as if* the nonhuman could give a linguistic account of itself, even though an important feature of its existence is its inability to give such an account. Simply creating the account distorts your understanding of the body for which the account is given. In discussing the interests and desires of a valley or a grove, you makes the valley or grove into a kind of person. Trying to speak for a body that cannot use language prevents one from genuinely understanding it. Personifying distortion arises in how an investigator of an ecosystem understands his activity.

You avoid personification if you aim simply to describe the processes that maintain an ecosystem or population's stability and how best to preserve it. But another recurring idea in environmental moral philosophy is that scientific approaches—or any mode of thinking that could be called rationalistic—are inherently dangerous, precisely because of its rationality. Scientific attitudes, according to Jim Cheney, impose human systems of understanding on nature, while true knowledge accounts for the expression of nature in its activity (Cheney, 2005, p. 109). The language of science abstracts from attitudes of love and care, which are essential means to recognize a body's intrinsic value. Scientific investigation carries a great danger of understanding nature as only a system to be measured, having no value of its own (Cheney, 2005, pp. 105–106). But an exhortation to nature's narrative and emotional elements indicates a limitation of understanding intrinsic value as a property. Scientific investigation aims to describe the properties of a body, one of which would be its intrinsic value. Yet the general consensus of philosophy holds that moral properties like value lie beyond the powers of science to learn.

I will explore environmental moral philosophy's tendency to suspicion or outright hostility to science in greater detail in subsequent chapters. For now, it is enough to note that the conflicts of narrative understanding and what environmental moral philosophy understands as the default scientific attitude creates these practical paradoxes in attempts to articulate positive political programs. The exhortation to give an account for those creatures who cannot speak, to conceive yourself as an advocate for animals, plants, and ecosystems that have been disenfranchised until now, is a recurring rhetorical image to motivate people morally. A narrative element of the ecosystem is required beyond what you can learn of an ecosystem's nonverbal aspects.

But only self-conscious creatures can construct narrative, because narrative is constituted in response to the question of "What happened?" (Deleuze and Guattari, 1987, pp. 193–194). Bodies incapable of asking this question are, therefore, incapable of genuinely having narrative aspects of their existence, unless they emerge from a relationship with self-conscious bodies like humans. An ecosystem can exist without such relationships. To treat narrative as essential to understand ecosystems themselves, and not ecosystems in relation with self-conscious bodies, improperly personifies ecosystems. Imposing this implication prevents successfully understanding a body in its singularity. The imaginative act required to create the account of a nonnarrative body smooths away essential features. "If the Great Bear River Refuge could speak, it would say the following," some representative account may begin. But the refuge cannot speak, and for a human to speak as if it could distorts its singularity, which we are trying to understand. If any action a human takes to understand a nonhuman body distorts that body, then the only first step to a solution appears to be that the human understander must be completely passive before the nonhuman body he seeks to understand.

Intuiting Intrinsic Value: Passive Knowing

Intuition has become a pivotal rhetorical device to recent moral philosophy. To make a potentially contentious idea more palatable to readers, a writer will begin her essay with appeals to common sense beliefs about what is morally right, then build an argument for her controversial conclusion from those premises. Agar's work is one clear example of this in environmental philosophy because his argument for a biocentric moral philosophy hinges on the common sense intuition that a body's moral value depends on whether it has a mind (Agar, 2001, pp. 25–29). His argument that all individual organisms have a mind is itself radical, but its root is an appeal to a common intuition. Virtually all philosophical advocacy for nature having an intrinsic value relies on such intuition. Throughout contemporary moral philosophy, the source of intuition is contemplation on an abstract problem, philosophical question, hypothetical scenario, or actual experience. Environmental philosophy's intuitions tend to focus on the worldly experience of nature, where we, in simple terms, feel the intrinsic value of nature. The environmentalist intuition is passively coming to know the intrinsic value of a natural feature as you actively explore it, such as hiking on Mount Robson. This is the sense in which I use the term "intuition" throughout the book.

I see two ways of understanding passive knowing in environmental philosophy: observation and intuition. Observation has garnered the least attention in scholarship, even though Aldo Leopold, a foundational writer in the discipline, offers the paradigm. If we are looking for some property of a body, whether that body is a single animal or an entire ecosystem considered as a unity, we will find it through detailed and meticulous observation of that body. Take, for example, his account of the lagoons at the Colorado River delta, which he explored with his brother in 1922:

> The still waters were of a deep emerald hue, colored by algae, I suppose, but no less green for all that. A verdant wall of mesquite and willow separated the

> channel from the thorny desert beyond. At each bend we saw egrets standing in the pools ahead, each white statue matched by its white reflection. Fleets of cormorants drove their black prows in quest of skittering mullets; avocets, willets, and yellow-legs dozed one-legged on the bars; mallards, widgeons, and teal sprang skyward in alarm. . . . Often we came upon a bobcat, flattened to some half-immersed driftwood log, paw poised for mullet. Families of raccoons waded the shallows, munching water beetles. Coyotes watched us from inland knolls, waiting to resume their breakfast of mesquite beans, varied, I suppose, by an occasional crippled shore bird, duck, or quail. At every shallow ford were tracks of burro deer. (Leopold, 1970, p. 151)

In this and other passages of *A Sand County Almanac*, Leopold displays a keen eye for large and small details of an ecosystem, all significant. A more urban person might observe green waters, and a variety of birds and plants, while unable to identify their species; he would likely not notice the stealthy animals like raccoons or lynx, while insects and deer tracks would certainly escape their attention altogether. In describing the coyotes and their food sources, Leopold also observes relations among the individual bodies of the ecosystem in addition to the bodies in isolation: the relations that transform plants and animals into food. Careful, detailed observation of bodies and the relations among them exemplifies an attitude of mindfulness. Mindful observation discovers the singularity of what is observed. A morality operating through mindfulness produces knowledge of what must be done in each unique situation, preserving in your thought the difference of this situation from all others (Smith, 2001, pp. 15–16).

When you observe nature well, its intrinsic value supposedly impresses itself upon you. Yet we can observe without understanding the value of what is observed, a failure to perceive the value property. Many passages throughout *A Sand County Almanac* are physical descriptions of some phenomena in the wilderness. Only the poetic invocations of the language itself communicates the value of this wilderness, the intrinsic value of nature. Attention to detail and scientific background knowledge are therefore insufficient to guarantee the intuition in which the property of nature's value impresses itself upon you. Leopold himself often needlessly killed wolves when he was younger (Leopold, 1970, p. 138). Though he would regret his actions later in life, he did not at the time perceive the value property of the wolves he killed, despite his detailed observational skills. You would therefore think that a priority of environmentalist philosophy would be investigating what perceptual and observational skills are required to guarantee the intuition of nature's intrinsic value.

However, descriptions of the intuitive experience tend to focus on the subjective feeling the intuition produces, and the resulting change in attitude toward nature, instead of the property of value that imposes this passive knowledge on the observer. A body's value property itself remains mysterious. Roman Briggs describes a simple realization as he remembers an adolescence working on the family farm, "coming to recognize the complexity and elegance which inheres within the land." He considers this realization a signpost of moral maturity through which he comes to love the land (Briggs, 2009, p. 168). Holmes Rolston III's everyday life growing accustomed to the rhythms and vistas of his rural community in the Rocky Mountains is enough for this property to impress itself upon him (Rolston, 2008, pp. 5–20). Rolston also describes the experience of hunting as revealing the value of the wild nature that a human hunter helps to constitute (Plumwood, 2000, p. 312). Jim Cheney describes how "a backpacking friend came upon a huge granite rock that expressed the more-than-human nature of the world in a particularly concentrated way that he could only refer to — with a certain awe — as 'sacred'" (Cheney, 1998, p. 276).

Each of these examples is quite different from each other, although they can all be considered, after the cliché, cases of people in touch with nature: farming as a boy, everyday rural life, the intensity of hunting, hiking through wilderness. Through their experiences, each writer comes to understand nature as more than a mere resource, or a place to live, or an annoying summer job. These experiences of nature provoke a reverential, respectful, loving attitude. Descriptions of the intuitive experience focus on the details of that experience, and the change in attitude it provokes. The intuitive experience is a rupture that forces you to perceive the intrinsic value of bodies. The narrative of your life is irrevocably changed, and cannot continue as it had. Your character and life priorities are reoriented toward an environmentalist virtuousness. But not all who experience nature experience this rupture of enlightenment. Another frequent theme of environmental philosophy is disparaging those who treat nature instrumentally, as a mere resource, something to be used or cleared away for the useful. You can "recognize the complexity and elegance" of nature, but find that complexity a repugnant chaos, and think a carefully manicured classical garden would be more elegant. Two people may experience the same body, but only one may be properly receptive to the value property's impress. Those who have developed an environmentalist's attitude are open to this impression, the proof of their openness being the development of the attitude. Perhaps we can understand the intuition's passive knowing by analyzing the attitude and working back to their catalyst.

Motives of Awe and Duty

When the environmentalist intuition produces an attitude of reverence or awe of what is utterly alien to humanity, philosophers tend to use religious traditions to uncover its causes. Not all analysts of religion in environmental moral philosophy are intuitionists, but they share with intuitionists a frequent focus on reverential attitudes to nature. Specific references are often made to Asian and the world's indigenous religions. These religions are supposed to encourage an attitude of worshipfulness toward nature, understanding nature as sacred and humanity as part of nature. This contrasts with the monotheistic traditions, whose creation myths supposedly condition their believers to consider themselves as having dominion over nature (Attfield, 2009, pp. 31–50).

Erazim Kohák describes an experience of nature in which "the presence of God is so utterly basic, the one theme never absent from all the many configurations of life's rhythm. The most basic trait of the world that confronts a dweller in the radical brackets of the forest clearing is that it is God's world, not 'man's,' and that here God is never far" (Kohák, 1984, p. 182). Kim Sung-Hae describes lessons from the *Xiang'er Commentary on the Daodejing* on living in the most passive way possible: a literal stasis of action, the value of nature is here understood as that which must not be disrupted (Kim, 2008, pp. 139–143). The thought of ancient Korea's Indramang community is said to provoke an intuition that nature must be revered (Lee, 2008, 115–134). Even ancient Chinese sex rituals have been advocated as a way of intuiting nature's value in a religious context (LaChappelle, 1984, pp. 102–121).

One problem with appropriating religious traditions for a political and philosophical environmentalism is that it often oversimplifies a complex ideological situation. There is a danger that the philosophical use of these religious worldviews and their holistic ethical philosophies tend to romanticize and essentialize complex cultures. The political power of religions motivate environmentalism's drive to co-opt them into the cause. Many appeal to religions because they already have a large number of devout and casual adherents around the world, so familiar as to be taken for granted. If you are persuaded that the religion you already accept mandates environmentalism, then accompanying activism and lifestyle choices more easily follow. Environmental philosophy's practical goal is to provoke a mass movement for conserving nature, and where a mainstream religion can be used to inspire lifestyles of environmental reverence and stewardship, such inspiration is welcome.

A religious approach is intended to inspire reverential attitudes, so to understand the personal power of awe, you must understand what it is to revere. Mountains are often objects of reverence in environmental philosophy, and Leopold's poetic injunction to think like a mountain means to take a point of view that more widely encompasses spaces, places, times, and possibilities of life and cohabitation than the perspective of a self-centred human (Leopold, 1970, pp. 137–140). Leopold, Rolston, Cheney, and Næss all take inspiration from their solitary mountaineering experiences. Reverence is an attitude constituted from an experience of sublimity, when words fail us in the attempt to describe the immensity of mountains compared to humanity. Næss can only say, with some irony, that a mountain can best be described as "big, very big" (Næss, 2005, vol. 10, p. 365).

Rolston makes this epistemic gulf between us and what we revere central to his theory of the foundation of environmentalist morality.

For him, mountaineering opens you to discover facts massively larger than what humanity's scale of action could possibly be: the truths of nature need not even fit the usual demands of logical consistency and coherence. Following Hilary Putnam's epistemic pragmatism, Rolston dismisses consistency and coherence as habits of human psychology to which the universe beyond a human skull need have no allegiance (Rolston, 2005, 137–174). A mountain's incommensurability with ordinary human thought is a power to inspire reverence through radical difference that forces upon human understanding a task beyond its capacity. Transcendence of the human capacity to understand would therefore be the property of an object that provokes the passive knowledge that nature is intrinsically valuable and should be revered. But this emerges from a relationship with humans, and so depends on humanity, which makes it vulnerable to human shortsighted misunderstanding and corruption. The environmentalist standards with which I began this inquiry cannot accept this.

Precisely identifying the property of nature and natural bodies that grounds their value free from human valuation continues to come up empty. Even Næss, an uncompromising advocate for understanding intrinsic value as a property, admits to having never identified the property itself: "I myself trust the intensively meaningful and spontaneous experience of the value of life in a free nature" (Næss, 2005, vol. 10, pp. 157–158). What matters is not a body's property of value, but the attitude of reverence and love that the experience of "life in a free nature" engenders. Yet without knowledge of the property that grounds the attitude, the attitude is illegitimate save for humanity's practical need to avoid ecological catastrophe, environmentalism for humanity's sake alone.

Processes that begin with a human action or thought are called anthropogenic, and significant quarters of environmental moral philosophy suspect such origins because of a danger that the thinker prioritizes human concerns over those of others.[2] Because the valuation understanding of intrinsic value would have nature's value arise from human thought, it cannot escape the danger of accidental reduction to instrumentalism. To avoid grounding nature's valuation on human decision, a body must possess a property of value that could impress itself on a passive human observer as the foundation of that observer's act of valuation. Yet any search for such a property falls back on human experience: you come to value a body, but cannot say why (Aikin, 2008, p. 29). That we must revere nature becomes an unjustified and unjustifiable dogma of the environmental philosopher. If you do not feel awe before nature, then you are not perceiving properly, sick with confused anthropocentric Western attitudes of hubris (Aikin, 2008, pp. 25–26).

Stewardship moral philosophy of environmentalism is an example that best illustrates the danger of slippage to which anthropogenic foundations of nature's value are vulnerable to human hubris. The image of humanity as the Earth's steward is sometimes justified in the context of Christian theology (Marangudakis, 2008, p. 245), although stewardship morality's moral justification need not rely on Christianity. Rachel Carson best articulates the moral concept of environmental stewardship. Stewards, for example, practice agriculture for the needs of their community. But they also practice ecological and biological science to understand how their entire ecosystem functions, and use this knowledge to develop an agriculture in harmony with all the elements of a mutually constitutive complex ecosystem. Essential to the role of the steward is that humans exercise responsibility for Earth (Bekoff and Nystrom, 2004, pp. 875–876). It is not the administration of a dictator, but the duty of a proud custodian who believes that the harmony his work maintains would fall to pieces without him.

Environmental stewardship programs, in practice, construct public moralities that can be highly effective in preserving ecosystems vulnerable to the destructive activities of human communities. Such moral attitudes and principles encourage mindfulness and care for nonhuman assemblages. Care emerges through the humble attitude derived from understanding the small scale of the human position in the universe. The epistemic imperative to make sense of the world creates feelings of inquisitiveness, which leads to mindful inquiry into how the world works and what must be done to protect it (Roach et al., 2006, pp. 58–60). Despite its practical effectiveness, a morality of environmental stewardship remains egotistical. The custodian's ethic can easily develop a self-concept as a benevolent guardian keeping all an ecosystem's elements in their proper places for overall stability. A human's peculiar powers of self-conscious thinking to create systems to understand the universe entitles a creature insignificant in scale to install itself as a kindly master over all. Being capable of systematic knowledge, a subject knows best, so directs and commands all an ecosystem's processes for what he takes to be their own good. The epistemic presuppositions of a stewardship morality result in a commanding attitude that irreconcilably clashes with the humble and sympathetic moral attitudes, which ground stewardship morality.

A notion from Isaac Asimov's *Foundation and Earth* illustrates the danger of stewardship morality's paradoxical attitudes of humble directorship. In the preceding novel in Asimov's series, the novel's protagonist, Golan Trevize, through means of unimportant and complicated technobabble, is able to choose in one moment the future evolutionary development of the Milky Way Galaxy as a whole. He spends this entire novel working out why he chose as he did, tortured by remorse over whether he made the correct decision, and whether one person should even have been empowered to do so. He eventually discovers the figure whose telepathic influence had actually been shaping the development of the galaxy for the last 20,000 years: R Daneel Olivaw, the last living robot. Olivaw justifies his mental manipulation of trillions of humans across the galaxy's inhabited planets with a principle of stewardship: Asimov's Zeroth Law of Robotics, "A robot may not harm humanity, or through inaction allow humanity to come to harm."

A single robot, insignificant in terms of physical scale, has a moral principle literally hardwired into him, which dedicated him to manipulate a massive population of people. Olivaw sees nothing questionable about secretly directing the development of a galaxy for 20,000 years, because his mental powers were greater than any other entity in it. Knowing that he knew best, he took control. His control was secret, and so unable to be critiqued. He would not have understood why anyone would critique him anyway, because he knew that he knew best (Asimov, 1986, pp. 346–349). The steward of humanity's development—or in my analysis, Earth's ecological development—is an authority figure beyond accountability to those over whom he has power. A human steward of an ecosystem takes authority to direct that ecosystem, knowing what is best thanks to ecological science and his own attentive observations. No matter how much material benefit he may actually provide to the ecosystems in his care, stewardship cannot escape this authoritarian streak. A benevolent dictator is still a dictator.

The Inadequacy of Intuition

The inescapable problem of understanding intrinsic value as a property is that we can never identify precisely what that property is, and how to guarantee that an experience of nature will induce an intuition of this value. The goal is admirable: to find that nature's value is a discoverable fact, that the value of a mountain is just as real, and just as foolish to deny, as its location, height, and mineral composition. The value of nature is part of nature itself. But not everyone who experiences nature experiences its value, so those who do experience its value must intuit some quality in addition to the perceptual experience alone. Many arguments for complex moral theories begin from premises their advocates declare intuitively true and adding the intrinsic value of nature to the list of these intuitive truths would be progress for conventional environmental philosophy. Yet the nature of this value often remains ambiguous or utterly mysterious. Despite this, Næss calls direct intuition "the most important category of knowing" (Rothenberg, 1993, p. 93), and distrusts the trend among environmentalist theorists of building complex new systems of rights, obligations, and responsibilities among individuals, species, phyla, and ecosystems. A philosophical system deriving moral conclusions from complicated arguments connecting a plethora of abstract principles has doubtful plausibility, says Næss, because it lacks "a clear intuitive basis" (Næss, 2005, vol. 10, p. 98). Such a clear intuitive basis must ground a theory whose purpose is to transform human civilization along ecologically sustainable paths.

For a purpose of such gravity, the intuition must be universal to the population, and (as the concept of intuition implies) it must be of a truth, therefore the same for all people. But recent work on intuition has uncovered serious problems for the role that environmental moral philosophers ask it to play. Intuitions act as premises in philosophical arguments. When an argument aims for universal scope, the premises must be uncontroversially universally true. But there can be no proposition that, with sufficient gumption and creativity of the disputant, is universally, obviously true without exception (Williamson, 2007, pp. 210–214). Intuitions regarding the same philosophical problems vary from person to person. Research on this variation has focused on problems in analytic epistemology, but we can apply these conclusions to the environmentalist intuition of nature's intrinsic value, taking due account of those who have the experience without the intuition (Alexander et al., 2008, pp. 138–155). This empirical research demonstrates that intuitions do not reveal obvious truths (Weinberg, 2007, p. 337).

Such variation in what people find intuitively true implies that the feelings of truth accompanying intuitions are signs of confirmation bias, or simply a presumption of obviousness that is taken for granted. An advocate of nature's intrinsic value would "trust the intensively meaningful and spontaneous experience" to justify that he had intuited an indisputable universal truth. Those who do not intuit such an indisputable truth would simply be wrong. At the very least, they would not belong to the community of environmental moral philosophers, because they do not share the premise that nature is intrinsically valuable apart from its possibilities as resources for humans. Such people would be part of the problem that those who do experience the intuition are trying to solve. But intuiting nature's value in spontaneous experience does not make someone into an environmentalist; it is simply the confirmation bias of someone who is already an environmentalist. Far from accessing an ordinarily hidden truth about nature, so-called intuitions of nature's intrinsic value is just one more groundless presumption in a philosophical discourse where groundlessness of belief is the norm.

If the intuition of nature's intrinsic value is just another confirmation bias, then environmentalism has no more profound import than any other political goal. There are clear pragmatic reasons for environmentalism. But the traditions of environmental moral philosophy I have discussed want their political program to have a deeper ground than expediency. Human survival is an anthropogenic motive, and so inherently suspicious. Yet the environmentalist intuition of nature's intrinsic value cannot find such deeper ground. Environmentalist political solidarity here takes on the unsavory tone of most calls for party unity. You can only participate in environmental moral and political debates if you share the foundational premises, such as nature having intrinsic value. Those who would question that premise are not members of the environmental philosophy community, because you are only a member of that community if you see no need to question that premise. This exclusionary social dynamic is how a topic of investigation, such as the intrinsic value of nature, becomes, in Scott Aikin's pejorative words, a dogma. Because intuitions cannot be genuinely universal, no intuition can ground a universal moral principle. The only function the property understanding of intrinsic value can claim is to facilitate social exclusion from discourse.

The failure of the property understanding to find an objective ground forces us to revisit the previously maligned conception of nature's value as emerging from the interaction of humanity and nature. If humans are to care for that which is not human, it will be a human decision to value, not a human discovery of value independent of humanity's relationship with nature. The human origin of the act of valuation is only a problem if all human decisions inevitably reduce to caprice and naive self-interest. But humans can investigate the world and make decisions paying mindful respect to the bodies that those decisions affect and involve, whether or not they are human. Respect for those bodies would be undercut if one understood the goal of an investigation as to speak for them. To speak explicitly for a creature implies that your words are what the creature itself would speak, had it the ability, even though not speaking is central to the creature's own identity. The moral exhortation to speak for the physically voiceless sounds like humble respect, but implicitly encourages the hubris of imposing narrative-formation capacities on bodies that do not have them.

If you believe that humans can investigate and learn from nature without exploitive motives or hubris and with curiosity and respect, then the act of valuation being a human act will not be a problem. The problem, instead, is to show that it is better for humans to have a respectful and caring attitude for the different, rather than a naively self-interested attitude. Despite its pretensions, the property understanding of intrinsic value fails to understand the different. We search for some property common to all bodies, some feature of all bodies in regard to which they are all the same: their value. A genuine attitude of care, however, need not rely on any commonality among what is cared for. The decision to care for something is a human act of valuation from which that thing's intrinsic value emerges: to value a body for what it is, its singular identity unique from every other body.

At this point, I can no longer refer to any intuitive clarity or truth to justify why singularity is the reason you should care for a body. Instead, the ultimate justification for believing that nature is intrinsically valuable lies in understanding humanity's existential dilemma of whether our existence is at all worthwhile. Accepting that humanity deserves to exist requires developing a network of concepts that justify a sense of wonder at the diversity of existence. My hypothesis is that intrinsically valuing singular identity itself is the best moral principle for such a goal. Diverse ecosystems preserve and encourage the generation of radical singularities, a new body being a significant departure from the precedence of those bodies from which it came. Accepting singularity as a moral value implies that one will value the diversity of bodies in the universe and will want to increase that diversity. Humanity is one species of that diverse whole.

To anticipate any critics, I will say that there is a very good reason not to value singularity: the problem of ubiquity. Because the identity of each body involves at least a slight variation from the identity of every other body, each body is a singularity. Change is the generation of variations from any given status quo, the production of singularities that did not previously exist. But if each body is a singularity, then valuing singularity means valuing all bodies absolutely. If every body that exists is singular, then it is no greater compliment to call a body singular than to say it exists. A property is worthless if all bodies have that property: saying everything is valuable practically amounts to saying that nothing is valuable. Callicott writes, "An ethic that embraces everything embraces nothing" (Callicott, 1989, p. 10). Camus writes, "To say that everything is privileged is tantamount to saying that everything is equivalent" (Camus, 2004, p. 529).

However, I cannot find such an argument convincing. First, consider an analogous point. Say for the sake of the moment's argument that string theory is the absolutely correct scientific and mathematical description of the universe (and lay aside any questions of scientific realism, as this is a single-purpose analogy only). This being the case, all the matter and energy of the universe is made of extremely small vibrating strings. To say that some body is made of strings is the same as saying it exists, so saying it is made of strings is an empty compliment. Yet one can still investigate strings scientifically and mathematically, eventually building a comprehensive understanding of the universe that is radically different than previous systems of physics. Analogously, if one accepts a principle of difference itself, singularity, being valuable, then one's entire approach to the universe is different from what it was before. Singularity is not just a hand-waving principle, as some theorists of the environmental intuition have treated that lucky insight. As the central principle of the ontology I explore in this book, it is a detailed concept that we can investigate.

We can conceive singularity in two senses, each better for some philosophical problems than the other. One sense, the least practically useful but the most ethically powerful, is the absolute sense: being unique, each body is valuable, period. This conception of singularity's universality alters no individual decisions; it is not strictly a moral concept. It functions instead as an ethical concept: an idea which, as it comes to inform your habitual thinking, transforms your character and personality. Valuation of singularity is not valuation by privilege. Universally applied privilege is, as Callicott and Camus say, contradictory, because privilege requires at least two classes of being: the exalted and the downtrodden. Valuing singularity universally does not directly impact individual decisions, but conditions the theme of one's approach to the world. Return to Sartre's *Nausea*, and Roquentin. His anxiety stems from an ontological problem. The singularity of the world is intolerable to him because genuine difference from what has come before implies the inevitable inadequacy of general statements. His only means of understanding the world is through general statements, which aim to be universally applicable over time. The validity of an act of understanding is grounded in the possibility of completely understanding the universe. If the world is not fully comprehensible through this means, then it is not worth bothering to try. This question, whether to bother thinking, is foundational for philosophy. I call it foundational because someone who has never thought philosophically before will be moved to do so from taking this question—Why bother?—seriously. As long as she continues to investigate this question on her own power, and not receive by rote any beliefs or ideas as dogma, she philosophizes.

The question "Why bother?" is not the only question that can spark a person's first philosophical exploration, but it is important for one of environmental philosophy's essential topics: survival. Even the most misanthropic environmentalist philosopher who believes humanity to be inherently destructive and evil, will consider survival a key goal. He simply hopes that nature will survive. Environmental philosophy arose with the political movement, and they share the goal of working to make sure Earth's ecosystems survive the destructive activity of contemporary industry. To embrace the environmentalist movement is to embrace this goal in the present moment. Survival, whether of humanity, nature, or some rearrangement of them in a new kind of relationship, is a goal. Accepting survival as a goal requires a positive answer to the question "Why bother?" Such an answer rejects the possibility of completion as a reason why we attempt to understand the world. If you consider environmentalism important, you desire that healthy ecosystems survive. This is not a nihilistic answer because there is no place in it for the universal indifference of someone like

Roquentin who cannot understand the value of existence. In some other time, place, and circumstance, some singularity may develop that would invalidate the concerns of today's environmentalist movement. Some singularity may one day develop that would invalidate even the most obvious and comforting truths of the present. The world may change in a way that would even require a negative answer to an existentially foundational question. Someone who accepts the value of singularity holds basically the same ontological view about the universe as Roquentin. She diverges from him ethically because she finds joy, not anxiety, in reality's constant escape from universal statements. Perhaps instead of Roquentin, Sartre's narrator would have been something like Poppy, the joyful primary school teacher from Mike Leigh's *Happy-Go-Lucky*.

The absolute conception of singularity can cause severe difficulties in an attempt to use it in practical reasoning, particularly the problem of destructive differences. From the fact of their singularity, each genocidal dictator, each human immunodeficiency virus (HIV), each sludge-spewing chemical plant is valuable, despite their destructiveness. However, thinking this way improperly articulates an ethical concept of difference as universal singularity, as if it were a moral concept for practical decision making. The moral or practical conception of singularity is essentially different from the absolute, although it is related. All bodies are singular, but their singularity can be mapped, measuring the ways and degrees in which a body diverges from a field in comparison to which it had been indistinguishable, or in comparison to otherwise similar bodies. We can measure the degree of creativity one singular body has, providing we keep a contrast class in mind, the conformity or similarities from which a body departs as it develops.

Generating a singularity is always a process of creation and destruction. The above examples of extremely destructive bodies throw the problem into its sharpest relief. Any change in a state of affairs destroys that state and its absolutely valuable singularity. But such change is process itself, the transformation of states of affairs. In transformation, what once was is destroyed. A new state is created, but destruction is required for that process of creation to succeed. Accepting the absolute value of singularity in a moral perspective focusing on the inevitability of destruction puts you in continual mourning. Ecology, with its focus on the interdependence of processes in their generation and transformation, supplies the basic framework of a guide to practical action: identifying which processes will empower or harm other processes (Spinoza, 1992, p. 207). Practicality and survival are different concerns from understanding singularity itself. A way to grow mindful of the power of your actions to transform the world is to investigate and learn to predict with reasonable accuracy what new singularities your actions will generate, and which your actions will destroy. Despite the creation and destruction in ceaseless change, valuing singularity in its absolute sense is how you appreciate the significance of change.

In its moral conception, singularity is a tool, a means to measure degrees and kinds of divergence. But the moral conception of singularity, in contrast to the ethical, does not imply goals for action; it is only a tool for understanding the consequences of your actions. I will describe the specific kinds of processes that are most important to measuring singularity in later chapters. What is most important now is to understand the partiality of practical, measurable singularity. Its implied goal is the preservation of your own body, and the preservation or production of other bodies that it is in your interest to preserve or produce. In moral contexts alone, how you should choose what bodies are beneficial to your interests is an open question. The choice could be made through any means, including caprice and profit. Considering singularity in an ethical context guides a personality to make a better practical choice. Such a personality focuses on the joy of producing novelty and divergence, and so in moral matters, proliferates diversity and complexity, growing the intensity of existing heterogeneity. This contrasts with the perspective that focuses on the loss that accompanies change. A nobler way to articulate the value of singularity ethically, in your character, is to exalt in the production of divergences, the diversification of existence, and the breakdown of stagnating orders to create new systems of relations and frameworks of existence. A recurring theme in environmental philosophy is that maintaining and increasing diversity and difference is an important good (Smith, 2001, p. 185). If valuing singularity means valuing diversity, then singularity makes an excellent candidate for the central concept of a comprehensive philosophical worldview for environmental thinking in the twenty-first century.

CHAPTER 2

LOOMING ECOFASCISMS IN THE VALUE OF NATURE

WHAT POPULAR OPPOSITION THERE IS TO ENVIRONMENTALISM TENDS to root itself in the same conceptual mistake: that if we care about nature, then it is to the detriment of the human race. This sentiment underlies the contention that backing away from resource development will cost people jobs and income, that reforming unsustainably high-yield farming and fishing practices will make people go hungry. Nature's loss is our gain, and to restrain human action in the name of environmental protection threatens and ultimately destroys humanity's freedom. Worse, to privilege nature over humanity would debase human rights and destroy the political liberties that the Western tradition of moral discourse has established and protected over the centuries. This perspective makes environmentalism a perverse philosophy that would sacrifice human children to save a whale, a fir, a flounder, or a valley. Ecocentrism and biocentrism, goes this critique, is a disguise for a new kind of fascism. However, this conception of environmentalism as ecofascism is based on a concept that, although popular and culturally intuitive, is false: that humanity and nature are ontologically separate: polar opposites in essence, any competition between which is a zero-sum contest. This is a mistake, and this chapter analyzes this mistake in terms of how we understand difference, and how difference can justify or obstruct our ability to care for a thing. If caring for the different means disadvantaging or debasing your own identity or what is the same, then a misanthropic conclusion is inevitable. A politically and socially viable ecocentric moral philosophy must begin from a conception of difference as singularity, difference conceived without any contrast class. One cares for a body because of what it is, its singular identity.

The Human/Nature Dichotomy

Even though its appearance in environmentalist rhetoric is practically ubiquitous, "dualism and the account of the self and of human identity as hyperseparated from nature" (Plumwood, 1993, p. 173) is the concept through which environmentalism as a political and social movement shoots itself in the foot. This absolute dichotomy of humanity and nature causes more problems in practice and thought than it repairs, and it repairs quite little. Consider what damage such dichotomous thinking can wreak on such an intuitively common sense idea like the precautionary principle for the avoidance of harm. If the action of some body would harm me, I would value some other body that interferes with the dangerous one to prevent its harming me. But a morality based on the precautionary principle alone, whether for the sake of an individual or community, keeps strictly human and personal interests and goals at the heart of its analysis. It therefore falls into the problem of anthropocentrism, nature's value being as an instrument to ensure human interests. Bodies are understood exclusively, or at least predominantly, as resources or hazards for those who value them. Nature is debased to the level of a human instrument.

A defining turn in environmentalist thought is problematizing instrumental justifications for care. Even so, there is no universal agreement that instrumentality is a problem. After all, the precautionary principle that motivates the practical goal of environmental thinking as the philosophy of a political movement is to aid humanity's survival by protecting the ecosystems in which we live. Human health crises in the urbanizing industrial period of human history, for example, provide a model for moral systems focused on ecosystem preservation and pollution avoidance (Preston and Corey, 2005, pp. 10–15). Contemporary environmentalist political movements have also understood the protection of the environment, the surroundings of a human community, as the motive of their political action (Kagawa-Fox, 2010, pp. 62–66). James Sterba argues that the political philosophies of egoism, libertarianism, and socialism all in their own ways, endorse environmentalism only as the defense of the human species. Every person, says Sterba, has license to damage or destroy what will harm it. This is a "principle of preservation," the precautionary principle under a different name, and given an imperative of active self-preservation, survival (Sterba, 2000, p. 273). But the moral imperative of survival here remains insular as humanity and nature remain alienated from each other: only humanity has a morally relevant survival imperative, and nature is a means to humanity's continued existence.

An ecocentric morality can only function when the motives for our concern at least pay some service to the difference of its objects. The human cares for the nonhuman; cultural bodies care for natural bodies; the civilized man cares for the wilderness. Yet most ecocentric moral philosophies still think with an absolute dichotomy of humanity and nature. I care for Not-I or Other because I care for what is different from me. Care is conceived through contrast. Conceiving of humanity and nature as two poles of an absolute dichotomy enables the growth of several other assumptions about what humanity is, and what nature is, which are not always friendly to humanity. If difference is understood contrastively, then care for what is different from me is care for the negation of myself. In such dualistic thinking, humanity's very essence alienates us from nature. Even though humans are part of the same evolutionary continuum as all other forms of life, humanity has transcended nature through its peculiar intelligence (Evernden, 1985, p. 118). Nature is understood as pure, peaceful, harmonious, an equilibrium. In contrast is a corrupt, violent, disruptive, humanity that is prone to provoke catastrophe. This unbridgeable divide between technological humanity and the natural is often taken for granted in environmentalist thinking (Kohák, 1984, pp. 32–34). This dichotomy is so prevalent in environmentalist rhetoric that it precipitates a critique of environmentalism as so hostile to humanity that it slides inevitably into ecofascism.

Environmental philosophy's grand absolute dichotomy is its I/Not-I of the human and nonhuman, culture and nature, or civilization and wilderness. Much of the political rhetoric of environmental activism begins from the image of a primordial balance of nature that human technology has disrupted, the goal of activism being to restore that balance. Humanity has interfered with the natural, and therefore harmonious, development of the Earth, and this disruption of nature's balance must be restored (Callicott, 2001, p. 79). Among those who accept the absolute dichotomy, the difference is understood in two ways: (1) The separation is inescapable, because part of the essence of humanity itself is to be unnatural; or (2) The separation has developed through history, usually through some human activity such as agriculture, urbanism, or industry. The absolute dichotomy is the dominant paradigm of Western thinking about nature: that humanity constitutes an absolutely separate and different order of being from all else on Earth, and is the superior member (Sessions, 1984, p. 29). The goal of environmental activism would therefore seek the inversion of this rank.

Spinoza was the earliest thinker to understand the absolute dichotomy when he writes of the opinion, widespread even in his preindustrial era, that humanity is essentially unnatural. "They appear to go so far as to conceive man in Nature as a kingdom within a kingdom. They believe that he disturbs rather than follows Nature's order" (Spinoza, 1992, p. 102).[1] Dichotomous environmental activism holds that because the root cause of contemporary ecological catastrophes is humanity's sense of superiority over the nonhuman or nature, to halt this catastrophic movement requires reversing humanity's conceit. Nature, in its balance and divine harmony, would be the superior member of the dichotomy. Humanity is therefore a disruptor, a corrupting force, an aberration from, and a threat to nature's harmony. Consider the following popular political debates: over the genetic engineering of plants and animals for agricultural and medical purposes; among organic, free range, and factory farming techniques; and how we understand the human-assisted evolution of domestic and agricultural animals, making pets and livestock out of wild creatures. The environmentalist side in all of these debates includes an unquestioned assumption that to be natural is to be good, and that human interference in a body is a corrupting influence that renders it unnatural (Siipi, 2008, pp. 71–103).

The political environmentalist demands that we must return to the basic goodness of nature, a stubborn dualism that reappears even in more subtle analyses of our planet's contemporary ecological problems. George Sessions understands "the diminishment of man and the diminishment of the planet and its nonhuman inhabitants as essentially one and the same problem" (Sessions, 1984, pp. 40–41). The beautiful notion that we're all in this together constitutes an essential nuance of thought. Yet Sessions maintains the absolute dichotomy of humanity and nature throughout the rest of his work. He simply carries out a less hostile inversion. For him, humanity is separate from nature to fulfill a role as nature's protector, a stewardship morality, which I have already explained as problematic (Skakoon, 2008, pp. 40–41). Even in philosophies such as Holmes Rolston III's, where humans are not essentially destructive, but technological life is itself essentially wrong. Those who follow nature do no harm, because nature is a state of being that does no harm to Earth (Kowalski, 2006, 165–183). Humanity is considered as having once been part of nature, but whose actions created this absolute separation, falling from a state of grace into nonnatural ways of existence, becoming technological, and therefore violent and chaotic. The structure of absolute dichotomy remains the same, but the concept now includes a dichotomy of nature's ordered harmony and humanity's disordered chaos.

The concept of nature and humanity's absolute dichotomy occurs persistently and pervasively throughout history. The ancient Greek concept of justice was understood as the harmonious order of nature, which human activity could either follow or disrupt. A central premise of William Paley's Victorian-era design argument for the existence of God is that nature is a perfect creation, which therefore could only have been created by a perfect being. The early conservationist movement in America also used the balance of nature as a concept in its rhetoric. In the early days of ecological science, ecologists conceived of ecosystems as superorganisms whose constituents act as organs maintaining a balanced harmony, although the consensus of modern ecologists has abandoned such strong ecosystemic unity (Jelinski, 2005, pp. 271–280). The Gaia Hypothesis in planetary ecology is a popular articulation of this concept of nature as perfect harmony. Nature, left to its own devices, will create the conditions that sustain itself and the planet as a whole. Only the unnatural activity of humanity throws off this balance and results in the contemporary global environmental catastrophe (Tobias, 1984, pp. 2–27).

The absolute dichotomy of harmonious nature and disruptive humanity permeates the American tradition of environmental philosophy, particularly its veneration of wilderness. According to J. Baird Callicott, the seminal works of Henry David Thoreau, John Muir, Aldo Leopold, and many others articulate the dichotomy through a romantic idealization of American wilderness as a real-life Eden, utterly alien to humanity and unstained by its corrupting influence (Callicott, 2008, p. 237). Ralph Waldo Emerson popularized an understanding of nature as an Eden that the merest touch of humanity would corrupt. Ironically, Frank Coleman identifies Emerson's dichotomous account of nature as encouraging wide-scale land theft and the clearing of forests. Wilderness was so alien to humanity that even value could not apply to it (Coleman, 2010, pp. 75–96). The poetry of John Haines, Alaska's legendary poet of wilderness, understands wilderness as the calm silence disrupted by the clattering noise of humanity (Browne, 2007, pp. 131–132). Erazim Kohák, a Czech emigré who settled in rural New England, also distinguishes wilderness and urbanity in terms of silence and noise (Kohák, 1984, pp. 36–37). Wilderness is, for Jim Cheney, a place that has remained exempt from humanity's fall into an instrumentalist way of thinking that craves only the exploitation of resources. Wilderness is where humanity can embrace a primitive way of life in perfect harmony that forgoes the abstractions of modern technological existences to rediscover our mythological ideal (Smith, 2001, p. 4). Wilderness is defined by a negation of all that is essential to contemporary humanity.

This American image of wilderness as Eden reveals how anthropocentric the absolute dichotomy truly is, because its framework conceives nature as the negation of humanity. Nature is not defined in its own terms, but only as a reversal of technological humanity, the mirror image of human nature. Only the idea of the corrupted has any positive content, while the purity of nature remains essentially mysterious, resulting in implausible Edenic idealizations of nature that have nothing to do with reality. Figures like Christopher McCandless and Timothy Treadwell demonstrate the folly of Edenic thinking, even while we may regard them admirably as exemplars who have given up hope for inevitably corrupt humanity to throw themselves into the purity of wilderness. They represent two visions of the Edenic understanding of wilderness, displayed for McCandless in the nonfiction novel *Into the Wild*, and for Treadwell in the documentary *Grizzly Man*. McCandless sought an authentic encounter with wilderness, a contest of immense danger whose authenticity derived from the genuine possibility of his death. Treadwell sought an even more radical break from human culture: literally to become a bear. Identifying humanity only with hate and cynicism, he saw in the bears he observed only love and sincerity, and committed so deeply to this vision that he even ignored the depth of violence in the actions of bears themselves (Drenthen, 2009, pp. 297–315). *Grizzly Man* director Werner Herzog offers an illuminating critique of the Edenic understanding of wilderness. Speaking from the Peruvian jungle during the making of his 1982 film *Fitzcarraldo*, he compares the jungle's continuous birdsong to the humming of constant murder, and shrieks of pain (Herzog, 1999). Violence is endemic to both humanity and nature. McCandless and Treadwell serve as cautionary tales about idealizing nature as the absolute other of humanity.

The absolute dichotomy of humanity and nature can even induce political apathy in an environmental activist. You become convinced that Earth has long passed the point of no return in the destruction of nature, because a purity corrupted by even the briefest exploration of a human or human tools can never be repaired. Humanity has voyaged over all the land surface of Earth and settled most of it; ships ply the oceans while manned and robotic submarines explore their lowest depths; airplanes and high-altitude balloons traverse the atmosphere at every level; Earth's upper atmosphere is filled with satellites and space junk; human-made robots have reached every planet in our solar system, flying by the outer gas giants and leaving litter on the surface of Earth's moon, Venus, and Mars. Humanity's violation of natural purity is no longer even restricted to the pedestrian scale. Biogenetic technology allows humans to meddle at the molecular level, creating through technological manipulation new kinds of organisms (Siipi, 2008, pp. 72–74). For someone thinking in terms of the absolute dichotomy of humanity and nature, the creation of microbial life forms with entirely synthetic genomes at the J. Craig Venter Institute would be the ultimate violation of nature's purity: the first completely unnatural life, unable even to fall from Edenic purity as humans have (Wade, 2010). One blogger described this event with the sarcastic title, "Mankind Creates First Synthetic Genome, Officially Replaces God" (Sancton, 2010).

If you understand nature as a purity that even the slightest human interference destroys, then once life is open to human manipulation at the molecular level, nature on Earth is entirely eradicated. Slavoj Zizek describes a world that is no longer nature, because all aspects of it are open to human violation of nature's harmonious equilibrium that had previously provided a comfortable home that barely needed our attention. "Nature is no longer 'natural,' the reliable 'dense' background of our lives; it now appears as a fragile mechanism, which, at any point, can explode in a catastrophic direction" (Zizek, 2008, p. 43). He describes a conception of nature that traditionally is the dense, unchanging background on which human dramas can play out without need for our reference or concern. Human powers having grown so great as to assault nature, nature is now an issue for humanity, either as a set of obstacles to be destroyed to access resources or as the threat of global environmental catastrophe. Nature has entered the human drama, but in doing so, it shows how misguided are the old beliefs about its passivity and simplicity as the mere setting of human concerns. Bearing the full scope and power of scientific and political attention to nature reveals an incredible complexity and agency of which humanity has been largely ignorant. The comfortable intuitions of the absolute dichotomy are no longer so certain.

Environmentalism's opponents nevertheless use the absolute dichotomy's framework to vilify the political movement as misanthropic and antidemocratic. If nature is all that humans and humanity are not and cannot be, then love of nature debases humanity. If nature is an absolute good, then humanity must be an absolute evil, an essentially violent agent of discord and corruption. And should not absolute evils be wiped from existence? I ask this question ironically, not only because I am a human who would like to continue living. Few state this explicitly, but a tone of disdain for humanity has recurred in Western environmentalist writing since Thoreau. This question displays the absurdity that can too easily follow from a belief in the absolute dichotomy of humanity and nature. If what you care for is essentially good, and it is also your universal opposite, then you must conceive of yourself as essentially evil.

There are three ways to critique this kind of environmental morality: (1) Show that nature is not uniformly peaceful, harmonious, and pure, an empirical investigation into the behavior of nonhumans at an individual and historical level; (2) Understand humanity as itself part of nature, albeit a part with some very peculiar capacities; and (3) Develop a justification for care that is rooted neither in contrastive difference nor in essential similarities. Each of these three approaches functions as a corollary of the other. If nature is not a pure harmony, then any destructive tendencies in technological humanity do not necessarily set humans apart from nature. If humanity is not different from nature, then it is not productive to justify care for nature on its being different from oneself. One's concept of difference is more nuanced than a starkly contrastive dichotomy.

Against the accusation that all environmentalist moralities are inherently misanthropic, all we need do is articulate an ecocentric morality that is not so. Yet even without the absolute dichotomy as a foundation concept, some ecocentric moralities would sacrifice too often the comfort of human interests and humans themselves for the preservation of nature. Such moralities take the ecosystem as a whole as the body of paramount concern. Individual animals, plants, fungi, and so on are the bodies that constitute the ecosystem, the parts of that larger whole, which Aldo Leopold calls a biotic community. Humanity can join such a community by settling within an ecosystem and acting in a manner that maintains the land's health, "the capacity of the land for self-renewal" (Leopold, 1970, p. 258). So the health of the land is, for such ecocentric moralities, the paramount good, and all human action must aim to maintain the land's self-producing power. Action that does not do so is evil and destructive, interfering with the processes of the land, which properly maintain vibrant varieties of life for an indefinite future.

While it achieves a complexity that most philosophies of the absolute dichotomy lack, such holist principles leave Leopold's philosophy open to attack as antidemocratic, more specifically ecofascist. Ecofascism outlines a political order in which individual welfare is subordinated to the welfare of the biotic community, the land, or the ecosystem as a whole (Schrader-Frechette, 1996, p. 63). On this interpretation of Leopold, such philosophy dispenses with moral principles of rights and dignity specific to humanity "because humans in the theory are only valuable insofar as they serve the ends of the biotic whole" (Starkey, 2007, p. 153). No essential element of humanity prevents humans from aiding the health of a biotic community. But the imperative to maintain ecosystemic health trumps any human's individual rights to do as they please. The early writings of J. Baird Callicott, Leopold's most prominent contemporary advocate, endorsed this misanthropic interpretation of land ethic philosophy (Zimmerman, 1995, pp. 227–230). A healthy ecosystem is a wild and varied place constituted through many different kinds of organisms together in relationships that benefit the whole and produce a stable ecosystem where that constitutive activity can continue. If a land does not maintain itself, it is because of human interference. An unhealthy ecosystem is one that humans have shaped according to a technological plan, where the only activities that may take place are those that are of direct benefit to the human planners. Leopold gives the example of a monoculture farm (Leopold, 1970, pp. 111–115).

The Theatricality of Enormous Industry

The contemporary industrial farm can indeed be understood as a large-scale assault on an ecosystem, flattening the diversity of a land with a single technological program, the mass production of food for humans alone, whether directly or to feed humans' livestock. No natural processes any longer maintain the health of industrial farmland, which has become an entirely artificial system. Humans are fed through the uniform production processes of industrial farms, but at the cost of the complex harmoniously self-producing ecosystem those farms have replaced. The moral quandary lies in humans having opposed themselves to the good of many species, which are more at home in a complex ecosystem than on industrial farms, even though mass food production is required to sustain an enormous human population. The enormity of humanity requires an enormous human industry to supply and satisfy the species' daily needs and desires. Understanding the contemporary ecological, technological, and social phenomenon of enormous industry, requires understanding the rise and fall of civilizations throughout human history, and the peculiar scale and intensity of our modern situation.

Ecological degradation and destruction have been the usual result of organizing ourselves in large numbers throughout humanity's history. Overexploitation of the land was a significant factor in the economic decline and fall of the great empires of ancient Mesopotamia and India. For thousands of years, large human communities have altered their ecologies with agriculture and resource harvesting for food and manufacturing. Industrial power has always existed in human history; before the invention of steam power, industry simply operated at a low intensity. As human economies succeed, their population increases, further straining their surrounding ecology, until eventually reaching a tipping point. Human growth outpaces the processes by which their food and resource bases renew themselves. The land does not respond with sufficient speed to human demands, and the economy collapses, taking the civilization with it, as its human population disperses again (Chew, 2001, pp. 1–10). As life in the area becomes less intense, disrupted processes develop a new way of flourishing after the transformations of industry, at least until its human inhabitants grow ambitious again.

Archaeological investigation of how numerous civilizations have collapsed over thousands of years puts the lie to any image of ancient humanity possessing some harmonious relation with Earth that we have since lost. But the contemporary problem of environmental pollution differs from the ancient ecological crises in two important, related ways. One is that the scale of contemporary human society is literally global. The Mesopotamian cities of Ur and Lagash were the centers of urbanizing and farming processes that deforested and overgrazed the Fertile Crescent. As local production became unsustainable, trade with other Persian Gulf cities like Dilmun and as far east as Balochistan supplemented Mesopotamian weaknesses, but only forestalled the inevitable fall. Mesopotamians dispersed from their cities and returned to lifestyles with lower ecological impact while the Fertile Crescent's ecological processes restored that territory's richness and diversity (Chew, 2001, pp. 18–26). Almost all of modern humanity lives entangled in urban lifestyles that produce enormous amounts of pollution and waste. The economies of every country and continent are so inextricably integrated through trade and investment processes that humanity now constitutes a single massive civilization under a variety of territorial political and social orders. In the event of a civilizational collapse, there is nowhere for humanity to disperse.

The other difference between the current anthropogenic ecological crisis and previous ones is the theatricality of enormous industry and its pollution events. Pollution events are more politically galvanizing than their large scale alone can account. People's attitude toward nature as a resource to be exploited and a challenge to be conquered has not changed from the ancient to contemporary industrial epoch (Riis, 2011, pp. 110–112). Sing Chew traces historically the ecological causes of the decline of humanity's great empires. These ecologically degrading activities proceeded very slowly, over decades and generations, so went largely unnoticed. Surviving poetry and philosophy from ancient empires show that the blurry boundary of nature and culture was often an issue for them. But their purpose was literary contemplation rather than activism to change the industrial processes of their civilizations (Chew, 2001, pp. 158–167). The reason attitudes regarding current ecological degradation have been galvanized is the speed and visibility of individual ecological crises, core elements of the theatricality of contemporary industrial activities. Mass media communicates terrifying images of ecological destruction, incidents of catastrophe so large and which proceed at such a rapid pace that people are shocked into activism. Rachel Carson's *Silent Spring* was the first artwork in contemporary environmental activism to capture industry's theatricality. Her artful and emotional prose describes in heart-wrenching detail the destruction of birds on a massive scale from pesticide pollution. Activism focusing on theatricality is a politicized aesthetics that calls attention to how unmindful industrial activity transforms scenes of nature that one expects to be aesthetically beautiful into the ugly and absurd.

Activism since Carson has only improved in cataloging industry's theatricality. Journalists travel to the Pacific Trash Vortex, a porous island composed of plastic trash that accumulates in the middle of the ocean at the vortex of its major currents. This island is twice the size of Texas (Handwerk, 2009). The Aral Sea, once Earth's fourth-largest lake, has shrunk to 10 percent of its former size after 50 years of Soviet cotton farm irrigation from one of the two major rivers that feed it. On the former Aral coast, camera crews gather to see enormous ships abandoned in the desert (Precoda, 1991, pp. 109–114). Photographer Lu Guang has a continuing project on enormous industry in northern China and Mongolia. His landscapes are dominated by skies of brown-green smoke, and earth the color of rust. Waters flowing through a multistory high dam are a brackish, metallic brown. A woman sits on a small, cement island in the middle of a lake of bright green sewage and decaying garbage (Walker, 2010). Edward Burtynsky's documentary *Manufactured Landscapes* sees him walking through mounds of fragmented computer parts—e-waste—that are kilometers long and several stories high, seeping toxic heavy metals into the ground water of a village in Zhejiang province (Baichwal, 2006). The largest scales of enormous industry have an intense theatricality that dwarfs even these images. The best example is China's South-to-North Water Diversion Project: the radical transformation of the entire Yangtze river itself, diverting enormous amounts of water from it to the Yellow and Hai rivers through a system of canals built to modify existing tributaries. By the 2050s when this project is completed, the entire landscape of China will be transformed (Diamond, 2006).

A scene of ecological degradation or destruction is theatrical when it is so enormous or strange that its very existence challenges one's presuppositions about what can exist. Such an object must, literally, be perceived to be believed. It is an event whose existence challenges its perceivers to believe that it can exist, even as they perceive it. This is why photographs and films are such striking means of perceiving industrial theatricality. Photographs, whether taken from valleys among mountains of shattered computer parts or from a helicopter flying over a canal linking rivers that span a country, are more transparent than paintings and more comprehensive than descriptions. A photographer can compose her image, but the body itself must exist in order to be photographed. Honest photographs are evidence for the existence of what is photographed.

More than this, the photographer's image has a depth of detail that goes beyond the ability of a painter to include or a writer to describe. "Photographs become standard evidence for historical occurrences, and acquire a hidden political significance... They stir the viewer; he feels challenged by them in a new way" (Benjamin, 1968, p. 226). Humanity can create catastrophes on a level comparable to nature, their scale and absurdity arousing confusion and fear in people who try to understand them. Theatricality inverts sublimity. "If nature is to be judged by us dynamically as sublime, it must be represented as arousing fear... for the aesthetic power of judgment nature can count as a power, thus as dynamically sublime, only insofar as it is considered an object of fear" (Kant, 2000, pp. 143–144). Conceiving of the sublime on a Kantian model, a sublime experience of nature makes humans fearful because humanity is fragile compared to nature. In the era of enormous industry, "fragility has just changed sides," (Serres, 1995, p. 20) as humanity has the power to inspire fear that used to be the exclusive ability of nature.

When we experience and consider enormous industry's scale and the intensity of its destructive processes, feelings of fear and awe arise in us that are suited for the sublime. But enormous industry is equally absurd as it is fearful, and this absurdity makes it theatrical. The Soviet ship rusting in the Aral desert that was once the shore of a vast sea inspires as much cynical laughter as fear. Satellite images of the Pacific Trash Vortex, and even the silliness of its name, inspire not fear, but ironic disdain. As well, these films and images inspire pity for the birds who drown after falling through the plastic 'land' trying to walk on it. On the Kantian account of sublimity, what provokes the experience of the sublime is utterly alien from humanity. For many intuitionist environmental philosophers, the sublimity of nature at its most grand and fantastic inspires reverence. When humanity has the power to create the sublime, such experiences inspire emotions and thoughts more complex than awe: fear and amazement blend with surreal and bleak humor. We can dialogue with documentary images and films of these catastrophes just as we interpret and engage with an artwork. Documentary art is how people far from these catastrophes experience them, and the means by which activists stir the emotions in recruitment. Human ecological catastrophe is rendered as theater, so the effects of these grand follies of enormous industry cannot be sublime, but theatrical.

Is Environmentalism Totalitarian?

Many environmental philosophers advocate ending this destructive theatricality by reducing the physical scale of human civilization: stepping back from enormity. Steady state economics, a style of economic planning and management that aims for stability and stasis rather than continual growth, is one possible route (Daly, 1984, pp. 94–95). Instead of massive industrial farms, humans would gather their food from small family or community plots nestled in a diverse ecosystem among a panoply of cohabitants and competitors. The size of human families and communities would also have to shrink. Human population reduction is one of the eight points of the deep ecology platform: "The flourishing of human life is compatible with a substantial decrease of the human population. The flourishing of nonhuman life requires such a decrease" (Næss, 2005, vol. 10, p. 565). If the problem is genuinely as urgent as contemporary industrial theatricality suggests, then advocating for smaller family size and awaiting the generational pace of change may take too long. Anyone who wants to hold an environmental morality focusing on ecosystemic health must deal with the problem that great human misery may be required to restore Earth's ecosystems to a healthier state, and that misery for such a purpose would be good. When the good of nature results in great ill for humanity, environmental philosophy faces a fearsome problem.

A moral stand prioritizing the flourishing of nature above that of humanity is denounced as ecofascism, the sacrifice of individual claims to life and liberty for the sake of the whole. In traditional fascism, that whole is the nation. In ecofascism, that whole is the ecosystem. There are enough similarities between contemporary environmentalist political programs and the ideas of Nazi philosopher Walther Schoenichen to give one pause. Schoenichen endorsed strict controls on industrial pollution, the protection of wilderness areas and their isolation from human contact, and critiqued Judeo-Christian theology for promoting a vision of humanity as utterly separate from and superior to nature (Zimmerman, 1995, pp. 212–215). Schoenichen's philosophy is compatible with the paradigm of environmentalist thinkers such as Næss and Leopold. The only exceptions in Schoenichen's thought are his most blatantly racist ideas—a culture that is authentically natural for him when society is rooted in a mystical union of blood and soil—and when he praises the Third Reich as the first ecologically progressive state (Zimmerman, 1995, pp. 217–219).

That a Nazi writer pioneered central ideas of environmentalist philosophy makes modern environmentalism no more racist than similar roots do modern rocketry.[2] Genuinely essential for the tension of environmentalism and democracy is the question of what to do about our enormous industry. Insofar as a technological lifestyle function to make all bodies potential resources, technology alienates humanity from nature. This alienation is not an absolute separation, the creation of a new ontological category, but nevertheless causes a profound misunderstanding of what humanity can be. A relative consensus in environmental philosophy is that humanity's most authentic life is to integrate with nature as part of a biotic community acting to maintain ecosystemic health. Such a human community would be small enough that you could bicycle from one end to the other, live by subsistence farming, free range livestock agriculture, and work in basic manufacturing, mostly along craftsmanship and small factory models (Næss, 1989, pp. 144–146). These communities need not be embedded in wilderness, but should share common resources with neighboring creatures such as wolves and bears (Næss and Mysterud, 1987, pp. 22–34).

Such a lifestyle combines the idealized peasant's connectedness to Earth with the scientific attention to detail and to underlying physical causes of events, which in the preindustrial period of human development was explained through myth (Weil, 1952, pp. 43–44). Of course, such communities could not support a global human population as large as the current one, so any large-scale sociopolitical switch to this lifestyle would require a drastic population reduction. This is how humanity would live if we would give up technologically intense lifestyle with their tendency to individual caprice, and follow nature to maintain the harmony of ecosystems. Humanity has uprooted itself and made itself homeless, the destruction of natural harmonies throughout Earth being the practical result of this arrogance (Heidegger, 2008, pp. 241–243). We must therefore remember how to "fit into the stability of ecosystems, thereby acknowledging our dependence on spontaneous nature for sustenance" (Klenk, 2009, pp. 70–71).

If the above description is the only social and political conclusion for environmental moral thinking, then environmentalism essentially challenges every political system centered on human liberty. But this agrarian vision of humanity challenges the conception of liberty as the capacity to uproot oneself from cultural traditions and live according to one's own terms. This concept of liberty is the foundation of the humanist critique of environmental morality. Humanist freedom is a revolution against the determinism of natural instincts, articulating a capacity to build a new order through a radical break with the past, living according to reason and not instinct. If a human's rootedness in history, family, community, cultural traditions, and a natural place in a harmonious ecosystem is the good of environmental morality, then such a morality opposes humanist freedom (Ferry and Renaut, 1990, p. 4).

From the humanist perspective, humans are the world's only agents of radical change, making them ontological breaks from the natural order. The actions of such creatures can engender catastrophe, but for the sake of novelty and freedom, that risk is acceptable. The creative reason of humanity is a condition for the possibility of technology and its associated lifestyles. The greatest power of human reason consists in breaking with any tradition that would keep one rooted to historical precedent and ecological niche. Physical and intellectual discoveries require leaving this rootedness behind. Human reason, says the humanist, is a powerful force for change (Ferry, 1995, 93–94). If an environmental philosopher would say that ecological catastrophe is inevitable after humanity conceives of itself as radically different from nature, then a humanist would declare catastrophe a contingent, unfortunate screw-up, admirable ambitions gone awry. But such a disaster is worth the risk compared to the ultimate good of bolstering the freedom of human reason to make its radical creative breaks from its homes.

Actual ecofascism is an ideology too extreme for most environmental activists to hold, despite Nazis like Schoenichen sharing ideas with democratic philosophers and activists like Næss, Carson, and Marxist ecologist Murray Bookchin. Nazi environmental philosophy developed in a reactionary political context articulated by writers such as Ernst Jünger, who critiqued technology as a downfall from an authentic German life in harmony with nature.[3] Falling away from ecologically harmonious authenticity transformed singular humans into identical worker-soldiers (Zimmerman, 1990, pp. 35–36). But fictions of race were more important to the violence of the Nazi regime. The human relationship of harmony through rootedness in nature was an image of the authentic German soul, contrasted with the homeless, disconnected urban capitalism of degenerate races (Bratton, 1999, pp. 5–7). Contemporary environmental moral philosophy developed in the context of the democratic and radical left. Næss praises small, almost autarkically self-sufficient rural towns as the ideal human community to maintain ecological health and harmony. These ideal environmentalist communities are built on habits of anarchist mutual aid among humans, animal and plant neighbors, and their shared environment. But Næss qualifies even his praise of anarchism. He is conscious of the dangers small town life poses for innovative thought, where there is intense social pressure against deviance from traditional patterns of thought and behavior (Næss, 1989, pp. 144–146). Freedom to innovate in thought can be just as important for an ecocentric philosopher as a humanist. The play of concepts in the actual world is more complex than the simple dichotomy of environmentalist and humanist implies. It is a mistake to identify opposition to ecologically destructive technology with opposition to human liberty of thought.

HEIDEGGER AND THE DISASTER OF DUALIST THINKING

Nonetheless, opposition to the dangerous technology of enormous industry actually has often stood against the freedom to make a revolutionary new beginning like the innovations that made such technology possible. Two dovetailing critiques shape a strong alternative. Regarding our popular conception of environmentalist politics, humanity and nature are not essential adversaries locked in a zero-sum competition, as I have argued throughout this chapter. Creating a single general category in your thinking for all that is not human grotesquely oversimplifies the variety of existence, and naturally tends to slip into talk of stark oppositions between immutable essences. If we instead conceive of humanity as one species among a wide variety of organisms and ecosystems, then environmentalist compromise becomes possible. Environmentalism will inevitably fail if we conceive it as inherently hostile to humanity. A successful environmentalist critique is a practical opposition to contemporary enormous industry. Environmental philosophy has been left open to humanist critique thanks to the influence of Heideggerian ideas about the essence of technology. I deal with the purely philosophical approach first, because it will generate further details for my critique of the absolute dichotomy of humanity and nature: a conception of care as a creative process, a peculiarly human freedom that can encourage the diversity and vibrancy of ecosystems.

A major difficulty with bringing Martin Heidegger into any philosophical discussion is that the discussion rapidly becomes all about Heidegger. Environmental philosophy does not typically focus on an analysis of Heidegger's concepts, as in academic secondary material. But Heidegger is an informal influence on the field. Næss observes that a majority of the younger environmentalist thinkers and activists he met throughout his life take Heidegger as a framework on which to build their own critiques of contemporary technology (Næss, 1997, pp. 2–7). Paul Shepard writes that Heidegger had a unique insight into the essence of technology, providing the only critical framework through which environmental philosophers could develop a nonindustrial way of thinking (Shepard, 1984, pp. 206–212). Critiquing technology is a primary task of environmental philosophy. Practically speaking, this at minimum involves developing a sense of restraint regarding humanity's technological powers. Human powers have grown enormous, and a moral philosophy whose major questions focus only on interactions between individual people operates on too small a scale to have any meaningful insight into modern humanity's unprecedented power. Our philosophy must catch up with our powers. As the twentieth century's most profound critic of technology, Heidegger cannot be ignored.

I already described the basic points of Heidegger's alternative to technological life when I discussed visions of the ecologically sustainable rural human community, living in a harmonious relationship with the natural constituents of its ecosystem (Zimmerman, 1990, p. 71). Heidegger describes the technological way of thinking as understanding all bodies solely as resources, purposeless vessels whose only good is how they can be used for human plans and activities (Zimmerman, 1990, pp. 86–87). Environmentalists, and Heidegger, rightly reject this instrumentalist view of the world. When we discuss the social and political diagnosis of the problem of contemporary technology, the Heideggerian critique of technology is quite often the environmental philosopher's. Here again is a thought that gives pause, because Heidegger was a Nazi, a racist, and a warmonger. Luc Ferry's influential critique of radical left political movements, including environmentalism, is that all leftists who are not liberal humanists are totalitarians. Radical left movements take seriously that there are some political problems that require us to defer to the best interests of a community or ecosystem, to the relative detriment of an individual's freedom. But accepting that some political problems require compromise between values of ecosystemic health, community solidarity, and individual liberty is not to endorse totalitarian government (Bratton, 1999, pp. 14–15). Heidegger diverges from environmental philosophy, to his own detriment, in how he envisions the development of the technological attitude and how to overcome it. In these conceptions lie the reasons why Heidegger's ideas will only retard environmental philosophy's creativity in the present and the future.

The actual issue is how to understand the peculiarly human capacities of technology, reason, and freedom. Heidegger considered the proper relationship of humanity and nature to be belonging together. There is a difference of essence between the two, but it is complementary, not oppositional, and certainly not an absolute dichotomy. Human thinking, the power to build complex systems of understanding, constitutes a clearing in which beings reveal themselves. This clearing is a framework of understanding, which provides a context for the appearance of beings (Heidegger, 2008, p. 447). Human thinking does not constitute this clearing independently of being; humans are the conduit through which being reveals itself on Earth. Heidegger understands human history as this play of being as it reveals and conceals itself (Zimmerman, 1990, pp. 192–195). Technology is terrifying for him because an essential attitude of technological living forces beings to reveal themselves as fuel only, that which may be used. So the technological attitude conceals the unique singularity of beings in their complexity (Heidegger, 2008, p. 332). Singularity can only be revealed in an environment where people can live authentically, where beings can properly reveal themselves through humanity letting them be: people live in small communities, production is a matter of careful handicraft, and nature plays a benign role in the everyday life of the community (Zimmerman, 1990, p. 108).

Beyond his influential reflections on the nature of technology, I consider Heidegger's thought most important for environmentalist moral and political philosophy in developing a concept of human freedom. Enlightenment philosophy developed a concept of freedom as a radical break from the determinism of natural instinct and mechanism, which has become the modern humanist perspective (Ferry, 1995, pp. 93–94). The environmental philosopher considers the concept of human reason as a break from nature itself to be an act of extreme hubris, which enabled humanity to become an agent of global destruction, the technological architect of enormous industry. For Heidegger considers human hubris to have concealed the singularity of being itself through our ontological forgetfulness. Having forgotten being's singularity, humans falsely conceive themselves as free from its determinations and able to determine being (Zimmerman, 1990, pp. 264–265). Human forgetfulness is the condition of the hubris that motivates technological civilization to dominate the world.

Humanist philosophies conceive of nature as a realm of passive determinism. Subjectivity is the origin of the only active force in the world, the only force that, in breaking from determination by past causes, could create the genuinely new. Understood morally, those causes are historical, because human freedom breaks from the stagnation of long-established social traditions. Understood ontologically, human freedom is the power to break from the physical determinism of linear causality. The freedom of the human subject was the conceptual pivot on which multiple domains of philosophy aligned in the development of Enlightenment thinking. Freedom to escape nature's determination grounds a promise to create a wholly free world through the universal exercise of human reason. Being the only source of freedom in the universe let humanity convince itself that its proper role was to dominate Earth (Ferry and Renaut, 1990, pp. 57–59). For Heidegger, humanism distills hubris and false consciousness. Humanity, the proper conduit of being, lets our pride lead us astray to believe ourselves in charge. The ecological and humanitarian disasters we perpetuate are the side effects of ontological shortsightedness.

Being itself is the only agent in a Heideggerian ontology, while humanity and nature are passive figures, acting out the concealing and revealing play of being. This principle prevents a genuinely productive uptake of his ideas into environmental philosophy. At heart, environmental philosophy remains a program of political and social reform and revolution, itself a radical break from humanity's technological lifestyle. Discovering ecologically sustainable ways of life is not a matter of waiting for being to reveal itself in some new manner, waiting for an environmentalist messiah, the god that will save us. The underlying principle of any political and social activism is that we have the power to change our own lives. Yet this is not a power that puts humanity in any ontologically superior position. To think so would only repeat the hubris that first facilitated the development of enormous industry. Environmental philosophy requires a new kind of humility. No matter how powerful the human urge to conquer Earth may become, perfect control is impossible. As in the case of Timothy Treadwell, when someone tries to force his own morality, philosophy, and way of life onto a nonhuman plane, the result is disaster. Humanity's enormous industry is capable of actions that strain human epistemic powers. This strain is the tension between the sublime and the absurd that constitutes theatricality. Environmental morality admonishes us never to act without careful attention to our situation. We must always know what we are doing.

To Act with Knowledge

Ecological science is one of humanity's best routes to learn the consequences of our actions. An ecologist is a keen observer, a detective who investigates every relationship emerging from the activity of the region under study. She works out how one relationship or set of relationships encourages, prevents, or forestalls the development of another. Ecology is the science of mindfulness, and so is an exception from the technological view of the world as a collection of resources that the Heidegger-influenced tradition criticizes. To be mindful is to pay maximum attention to the world, your surroundings, thoughts, and projects. Mindfulness is thoughtful attention, not only focusing on the obvious features of your existence, but working out their deepest details, the subtlest connections among phenomena (Riggio, 2012, p. 402). A mindful observer thinks without disciplinary boundaries, because bodies do not let the divisions of research fields constrain what they affect. For example, facts of epidemiology or microbiology could have considerable impact on facts of botany. Ecology is a transdisciplinary knowledge, a science that pays attention to the problems of a world that does not easily fit into the definitions of scientific disciplines and subdisciplines. Ecology's transdisciplinarity offers the opportunity to transform philosophy, as philosophical analysis and critique integrate with ecological investigation and policy formation. Philosophy that takes ecology seriously leaves the isolation of an armchair philosopher's intuitions of nature's value to take an active role in investigating the world.

University of North Texas (UNT) philosophy department workshops on recovery from Hurricane Katrina and on the ecology of Cape Horn in Chile illustrate mindfully active knowledge. The Katrina project examines the links between failures of government administration, architecture, urban planning, and environmental neglect, which together contributed to the destruction of large swaths of New Orleans. Its transdisciplinary framework examines relationships that only become visible when several knowledge disciplines combine their investigative powers (Kates et al., 2006). UNT's role at Chile's national park at the Cape Horn archipelago uses the experience of studying ecosystems to work through environmental philosophy's problems and improve the study of ecosystems through philosophy's institutional critiques. "The approach taken by UNT Philosophy sees the goal of achieving policy relevance as being tied to developing new institutional types of knowledge production. To be specific: environmental philosophy must challenge the current institutional definition of what counts as philosophy" (Frodeman, 2008, pp. 604–605). Ecology studies the powers of bodies acting simultaneously together in the world to constitute the place in which they live. An ecological philosophy examines the possibilities and deeper meanings of these active physical relationships.

These relationships are often surprising, confounding the expectations of an amateur observer who has not grown accustomed to what ecology brings into focus, the inventiveness of creatures. An intuitive environmental philosopher may undergo a mystical experience of the value of nature by experiencing a gigantic oak tree. He may stand in awe of the tree's size, its singular beauty, and presumably its intrinsic value. He may consider the tree in the absolute value of its singularity. At the sight of that tree infected by a disease and dying, that intuitive philosopher might mourn its passing. But an ecologist sees much more. A dead tree, half-fallen, ensnared in the branches of its neighbors, is a shelter for raccoons and grouse, the latter also feeding on growths on the side of the oak. The activities of a swarm of bees fill a hollow oak with honeycombs. The layers of an oak's dead bark are an excellent place for insect eggs, larvae, and cocoons. Those insects also make a feast for chickadees. An ecologist, Aldo Leopold in this case, observes this vast multiplicity of productive relationships that the death of an oak tree makes possible (Leopold, 1970, pp. 78–82). These joint productions constitute the many practical values of the singular tree. The interpretation of Leopold's philosophy that emphasizes a holist concept of the biotic community, where individual claims and desires are dismissed to safeguard the integrity of the entire system at all costs, leaves him open to charges of ecofascism. But holist conceptions of ecosystems evoke an image of nature as a harmonious equilibrium, expressing the Edenic imagery of the absolute dichotomy. Leopold's accounts of how ecosystems actually operate describe freely acting bodies and systems in dynamic relationships of intimate interdependence.

An ecologist takes humanity to be one species among many to constitute ecosystems whose structures are dynamic. The apparent stability of an ecosystem's structure is more often than not a complex and messy collision of a huge variety of organic and inorganic bodies whose dynamic fluctuations have become so routinized that the ecosystem falsely appears to be a genuinely unchanging equilibirum to the untrained eye. An ecosystem is a roiling mess of cohabitants taking advantage of each other, as the chickadees, raccoons, and grouse take advantage of a collapsed tree. When philosophy takes the viewpoint of ecology, all bodies are understood to be active, their collisions constituting the incredible variety of nature. The dominance of one type of body creates the dull, easily collapsible monoculture of a mechanized industrial farm. But a dominant body like a monoculture farmer can also conceive of ways to restore variety and wildness to the boring and moribund. An *ecological* philosophy vilifies only the destructive effects of irresponsible creativity, action without mindful attention to the world where this creative force acts. Ecological understanding collapses any absolute dichotomy of humanity and nature without reducing either pole to the other. Focus on the relationships among bodies, how the activities of those bodies interact to constitute those relationships, and the possibilities those relationships open and close. A body has no essence other than its potential activity, what it can do. Ecology is the epistemic context in which one can best articulate the value of singularity.

Ecologically informed moralities are superior to a traditional humanist morality, in part, because there need be no zero-sum contests in ecological contexts. Ecological knowledge and understanding exposes the absolute dichotomy of humanity and nature as a juvenile simplification of a world that is far more complex, whose problems have more room for creative maneuver. Ecofascist or misanthropic interpretations of biocentric and ecocentric moralities depend on the absolute dichotomy to make sense, so accepting ecological knowledge neuters their critique. Humanity can thrive along with a vibrant, diverse nature, just not with enormous industry in its current character. Ecological thinking recognizes the complexity of bodies and systems, that they are composed of many aspects and constituted from many smaller and larger bodies that affect its generation.

Some of these bodies become physically enfolded in the body they generate, and some remain outside, providing the context of its generation, its assembly of itself. If we understand a body ecologically, then we know to distinguish the process of a body's generation from those bodies that constitute it. Each body is singular: it exists in a particular place, for a particular duration. Many bodies may share properties. For example, two bodies can have the same weight. But each body exists at its own place and time, each property of that body belonging to it alone. The equal weight of two bodies is a coincidental occurrence of the same measure for one aspect of two different bodies. Commonality is a coincidental resemblance of properties. Singularity is identity, the essence of a body existing in its uniqueness (Deleuze and Guattari, 1987, pp. 261–262). The central characteristic of an ecological morality is that "care requires attending to and respecting difference" (Kretz, 2009, p. 120). A body is what it is, and no other body can be exactly the same. Singularity is identity considered in itself, having passed all limits of commonality. Understanding care as justified through the singularity of that for which one cares removes the sting of anthropocentrism from the act of valuation. Through this framework, we can value a body intrinsically without having to depend on some specific value property. The body itself is valued, not some property of the body that can be isolated from its entire constitution and shared with all other value-worthy bodies. We value a whole body, all its properties, its situation of existence, its relationships with all other bodies, all of which is also why we value it. The moral act of valuation rests on a metaphysical principle of the singularity of a body's identity.

Chapter 3

Two Paradoxes of Practical Philosophy

Contemporary environmental philosophy began as a political movement. For this reason, Andrew Light distinguishes environmental philosophy from other branches of the discipline by saying that where the rest of philosophy seeks truth, environmental philosophers seek to craft good policy (Light, 1996, p. 327). Environmental philosophy took seriously the activists who would give their lives for the sake of a forest or to stop the hunting of wild animals. While there has been disagreement on whether environmental activism should use sabotage or other violent methods, theorists take activists' care for nonhumans to heart, and work to build a systematic philosophy for this concern (Nash, 1989, pp. 193–195). More than just the personal commitment of the activist motivates the best among environmental philosophers to break from mainstream tradition. It is commonly accepted in environmental philosophy that human industry caused a crisis unprecedented in scope and destructiveness. One often reads that the ability of Earth itself to maintain life as we understand it is at risk. Contemporary environmental philosophers see themselves as leading a charge to transform society radically. The ultimate goals may vary, with some advocating sustainable economic development and escaping dependence on fossil fuel, and some advocating a return to a low-population agrarian society. Each environmental philosopher has his or her own vision of the ecologically friendly civilization, but almost all call for radical change.

Given the deep integration of environmental philosophy's identity as a discipline with political activism and policy, it finds itself in a bind. The urgency surrounding environmentalist activism is largely motivated by the unprecedented enormity of contemporary industry. Philosophers were some of the first writers to become conscious of this urgency, but today they have largely fallen behind in activism, as more immediately practical professions foster vibrant environmentalist movements. Environmental philosophy's contemporary discourse tends to consist of academic arguments over the details of principles and systems of norms. The audience for these increasingly technical disputes is rarely the general public, but instead other specialists in environmental philosophy, who work in the university sector. As environmental philosophy has found itself a place in academic institutions, its discourse has disconnected from the environmental movement in politics.

Philosophy's Crisis of Public Relevance

The disconnect of an increasingly technical field of philosophical discourse not only applies to environmental theory, but to the discipline as a whole. Michèle Lamont, in her sociological studies of academic disciplines, declared philosophy a "problem field," because of the increasing insularity of its technical language and daily practice. Her interviews with practitioners of other humanities disciplines find that outsiders consider philosophical projects as "irrelevant, sterile, and self-indulgent," and philosophers themselves as occupying their own planet, separate from the rest of the humanities and the wider world. The technical languages of many philosophical subdisciplines are difficult for nonphilosophers to understand at face value, and she found few philosophers willing to make the effort to be understood by people, who do not already understand them. Lamont found that many philosophers commonly regard other fields with arrogance, and dismiss them as lacking rigor (Lamont, 2009, pp. 64–65).

Philosophy as a university discipline now faces the harsh material consequences of our insularity and arrogance. Lee McIntyre cites the near-closure and eventual downsizing of the philosophy department at University of Nevada-Las Vegas (UNLV) in 2011 as the first instance of a growing crisis in the field. UNLV administrators, faced with tightening budgets and an economic recession, cut funding and positions in their institution that were seen to produce no public benefit or having public relevance. Philosophy was judged to be a discipline whose products seem completely disconnected from the concerns of the general public, and was slated to be cut entirely from the university. After an uproar in the community of professors, the department was salvaged, though all junior-level faculty were let go. But the fact that a philosophy department, once taken for granted as a necessary element of a university, could face such drastic action is a sign that the discipline's over technical discourse and the community's resultant insularity does not protect it from serious danger (McIntyre, 2011).

Regarding environmental philosophy, we can take two responses to this problem. I believe both to be excellent ideas. One would be to try to resume the political lead again, which means joining a transdisciplinary conversation with farmers, economists, architects, and engineers. Environmental philosophers must overcome a prejudice that lingers in the field about such professions, because in the beginning years of environmental philosophy many such professionals were adamantly and universally opposed to environmentalism (Leopold, 1970, pp. 272–276). The second response is to keep back self-consciously from immediate political action. This response conceives philosophy as creating a comprehensive way of understanding the world that complements the moral debates on norms. A philosophy comprehensive of ontology, epistemology, ethics, and morality is relevant to political action, supplying new ways of thinking about the universe that constitute a conceptual framework for environmental activism. My own approach is more heavily flavored by the second response, but the transdisciplinary eye of the first guides my search for relevant material for thinking. From a philosophy like this, ordinary people who hold environmentalist values can build a more complex vision of humanity's place in the world.

The First Paradox: Unifying the Eternally Separate Thought and Action

Environmental philosophy began as a blend of philosophy, science, and activism. Seminal figures in the tradition such as Aldo Leopold, Rachel Carson, and Arne Næss continue to be referenced as forerunners and inspirations for both thinkers and activists. The first paradox of any philosophy that develops from a political movement is the notion that an esoteric network of concepts can provide, if not a program, then an impetus for political activism. Næss makes an illustration of this paradox. His concept of the intrinsic value of nature is a fruitful starting point. The theatrically destructive acts of enormous industry provoked a powerful sense of urgency that motivated him to create this concept. All facets of the environmental movement share this urgent fear that no matter what action we take to change industry and society from its destructive habits, it is already too late. This sense of urgency explains why Næss described the concept of intrinsic value in relatively simple terms; sometimes too simple. With the goal to make his theories applicable and expedient for mass movement, he articulated this fascinating concept as a mere dogma.

Næss is a tragic figure, wedged between the complexity of his own thinking and the urgent desire to simplify his ideas for political action. The blend of scientists, business people, artists, activists, and philosophers in the environmentalist movement constitutes incredible friction between the restraint asked of economic and technological development, philosophy's careful attention to argumentation, and the urgent provocation of the activist. Even so, real-life exemplars managed this friction relatively well. Consider Leopold, who worked as a professor at the University of Wisconsin-Madison, conducting the first scientific studies of Wisconsin's ecology, and writing essays intended to stir people into abandoning the popular image of nature as a resource to be scientifically managed. Ecocentrism, the moral philosophy that developed from Leopold's writings, is just one response to the radical situation of human activity causing ecological destruction (Callicott, 1989, pp. 3–4). But today's environmental activist understands her role as separate from a philosopher. Activism is the task of reporters, photographers, documentarians, and political organizations.

So the media advocacy of the works of Leopold, Næss, and John Muir is no longer a philosopher's task. What about designing the social and architectural structures of ecologically sustainable human communities? Næss laid out plans for such a society in the 1980s (Næss, 1989, pp. 138–146). Val Plumwood's late work includes complex analyses of what economic, political, and social structures create and accentuate ecological disasters today, and she outlines methods of reform along sustainable lines (Plumwood, 2002, pp. 81–96). Yet these plans are not just designed, but enacted by people who have little or nothing to do with the field of philosophy. I learned all the following examples from a *single* conference on sustainability that I attended in 2010. It was a typical conference, each presenter displaying a new environmentalist innovation. If this is a sign of the field's usual activity, then new initiatives and experiments in ecological health are widespread and prosperous.

Farmers are developing new methods of nonindustrial agriculture that can produce food on a large scale without overtaxing soil. They are also integrating insect colonies with fields, an essential task, given the disastrous effects of neonicotinoid pesticides on the bee populations required to fertilize agricultural plants (Reeve, 2010). Experimental farms in Oregon are building sewage and sanitation systems that recycle human manure into fertilizer with the goal of creating a zero-waste urban plumbing infrastructure (Dvorak, 2010). Professors of architecture are integrating ecological sustainability into their curriculum, creating a new generation of architects who design buildings, towns, and cities with a goal of zero waste production (Al-Hassan and Dudek, 2010). These architects are working now, in one case designing a radical method for household-based rainwater harvesting in India's most densely packed cities that could almost completely replace reservoir water supplies whose dams disrupt local rivers (Chipalkatty, 2010). Urban architects have already absorbed lessons of twentieth-century environmental philosophy, which describe an ecologically fruitful lifestyle as founded in understanding your life as interdependent with your place on Earth (Ingold, 2008, pp. 1801–1802). Architects now put these concepts to work designing urban habitats that integrate their human dwellers with surrounding ecologies, rather than bulldozing them (Derbyshire, 2010).

Environmentalist elements are growing in prominence in the fields of architecture and urban planning, one noteworthy institution being California Polytechnic's Center for Regenerative Studies. Pioneered by John Lyle, the center's philosophy conceives of architecture not as the production of static edifices, but as the management of cyclical ecological processes, whether in urban or rural environments (Orr, 2002, pp. 162–167). Ecological thinking has entered even accountancy, one example being a movement to abolish the "miscellaneous" column from reports, because it is often used to hide environmentally destructive costs or ignore ecologically fruitful activities (Parkinson and Chew, 2010). These ventures display a political environmentalist movement driven by practical scientists, business professionals, and tradespeople. Although work may not be proceeding with the speed that more militant environmentalists believe is required, work is getting done, driven by vocations that in Leopold's time were hostile to any environmentalist appeal (Leopold, 1970, pp. 265–272).

So where does this leave the philosopher in the transdisciplinary environmentalist political movement? Among all the professional disciplines in the modern environmentalist movement, the philosopher is the only one who cannot rest simply with believing in the rightness of his cause and action. A philosopher must both believe as well as understand how a belief is justified and motivates political activity. This is why the debate over whether we should value a healthy nature as a means to human flourishing or as having intrinsic value consumed so much philosophical energy. If a person ultimately acts for his own good exclusively, he would remain ethically suspect despite his beneficial effects (McShane, 2007, pp. 169–186). But activists carry on the environmentalist movement's political work, no matter the ultimate ground of their concern.

Academic environmental philosophy also ploughs a significant amount of energy into interpreting the writings of pivotal figures in the history of environmental thinking like Thoreau and Leopold, among many others. But problems such as, for example, what exactly Thoreau's concept of health was, do not matter to someone organizing an ecologically sustainable farming community (Casada da Rocha, 2009, pp. 201–215). Whether Leopold was influenced by philosophical pragmatism has nothing to do with the current task of working ecologists (Callicott, 2009, pp. 453–486).[1] The scholarly interpretation of sources, influences, and subtle technical details of early environmentalist thinkers likely matter little to contemporary projects to put human societies and economies in an ecologically sustainable path. However, such interpretive and historical exegesis is well suited for publication in academic specialty journals that are little read and often heavily access restricted anyway, but weigh heavily with a professor's tenure evaluation committee. Some philosophical debates and arguments develop complex systems of morality and ontology. Yet architects, engineers, farmers, and accountants are not waiting for philosophers to give them a theoretical framework for the moral beliefs they already hold deeply. They are already working, and philosophers are usually left behind.

Environmental activism began as a project of philosophy. It was unorthodox philosophy, sometimes written as diaries and travelogues, but in creating concepts for environmentalism, it was still philosophy. Perhaps the transdisciplinary writing of Leopold and Næss had to replace the aesthetic explorations of Thoreau and Muir. You can hardly learn how to fix an ecological problem, or repair relationships among human communities and their environments, if you do not learn how ecologies develop. Yet contemporary environmental philosophy continues a prominent trend of writing as if humanity still does not know the harm that modern enormous industry does. For example, Lisa Kretz writes in 2009 that "the failure of many humans to locate themselves ecologically has contributed directly to the current ecological crisis," as if there were not already a globally popular movement encouraging people to reconsider their lives in exactly this way (Kretz, 2009, p. 116).

Ecological concern is a matter of popular social consciousness. Not everyone radically disconnects their lives from industrial civilization. But everyone who followed news of the *Deepwater Horizon* oil well disaster or the Fukushima nuclear plant explosion has engaged with the issue enough to understand that human industrial activity does ecological harm. Academic environmental philosophy is hardly the vanguard of public advocacy and activism. The political lobbies of activist organizations like Greenpeace and the Sierra Club rarely include the philosophical community in their discussions. Philosophy's usual styles of discussion are in an important respect incompatible with the requirements of effective political activism, or with developing architectural and engineering projects for ecologically sustainable human communities. As evident in the near shutdown of UNLV's philosophy department and the widespread lack of respect other academic disciplines have for philosophy, few consider philosophy relevant to contemporary concerns anymore.

The Second Paradox: The Nuance of Truth and the Simplicity of Effectiveness

When you agonize, argue about, and painstakingly improve a system of morality, you hope that people will eventually follow that system. No moral system can control human action absolutely. Humanity has long had moral injunctions against killing people, yet there is no community where murders never take place. But today's environmental activists rarely wait for a moral philosopher to tell them if they are right before doing their own practical work. Environmental philosophy's debates over subtle distinctions and careful conceptual explorations do not fit easily into a political slogan such as "Not Man Apart!" (Jeffers, 1995, front matter) and "The Future Is in Our Hands!" (Næss, 1989, pp. 88–89).[2] Such sloganeering might strike a philosopher as simplistic and without nuance. Yet to many others, understanding a concept as a simple slogan can render an idea personally, socially, and politically meaningful, encouraging its uptake. It often shocks members of the intellectually sophisticated minority to see the power of a slogan when people incorporate it into their lives (Max, 2010).[3] You understand an idea better with attentive, nuanced thinking, but the second paradox of environmental philosophy's political heritage is that philosophical understanding stands in the way of direct political action. Subtle thinking generates differences of opinion among the members of a political movement, which grow into interpersonal tensions and breakdowns in the unity of mass action. The conception of intrinsic value as a property of natural bodies may not stand up to detailed philosophical scrutiny, but its simplicity makes it a brilliant slogan to motivate a political movement.

Næss gives the ontological concept of nature's intrinsic value a central political role without anything like my concentrated effort to overcome its shortcomings. His example shows how the disconnect of academic environmental philosophers from public politics permits a philosophically untenable concept to be politically effective. For Næss, the environmental philosopher engages in two tasks: the creation and refinement of philosophical concepts, and the translation of those concepts into effective political and social norms. The first task is a nuanced analysis that Næss called "precisation," an awkward English construction that literally translates a common Norwegian word. Næss developed the concept of precisation in his four-volume work *Interpretation and Preciseness*. "A sentence S_1 is more precise than another, S_0, if and only if the latter, S_0, permits (in ordinary or technical talk) all interpretations of the former, whereas the former, S_1, does not admit all interpretations of the latter, S_0" (Næss, 1989, p. 42). Each step in reformulating a sentence or description to narrow and restrict its possible interpretations is an act of precisation. A more precise articulation of an idea narrows the number of ways one can understand that idea, while that smaller set of more precise meanings does not add any interpretations that were not part of the idea's earlier, less precise articulation. Ideally, philosophical analysis is a precisation process. Beginning with a simple articulation of some idea, which might be understood in many different, equally plausible ways, its philosophical analysis reduces its number and variety of interpretations. Nuanced philosophical reasoning requires precisation because philosophical concepts and systems are so technical and complex, but the depth of detail in very precise accounts makes them difficult to understand in full.

As one makes an articulation more precise, it "is apt to be long and complicated" (Næss, 1989, p. 43). Sentences exhorting people to political action must be easy to understand.[4] "The choice of a rather indefinite and ambiguous sentence in the most elementary argumentations makes this fairly short and easily understandable and opens a variety of different possibilities for derivation and interpretation"

(Næss, 1989, p. 43). Short, slogan-like sentences are easily understood, but because they are not very precise they can be understood in a variety of ways, some of which may be mutually incompatible, or even contradictory. A statement in political activity can have as many divergent interpretations as there are listeners, as long as all agree that the same action plans follow. Simple sentences are not comprehensive accounts of reality, but "keys to the discovery of new sets of connections between things, the sum total of which may not at all be adequately described in terms of the original statement" (Næss, 1965, p. 133). Such simplicity, for Næss, is essential to spark the epiphanies through which abstract philosophical concepts motivate action, whether that action be further conceptual discovery or political agitation. Short exhortations drive people to the political actions required for a social movement like environmentalism.

One way Næss expressed this idea was through developing a system of logical notation in which normative propositions–exhortations–are imperatives ending in exclamation marks. Such imperative statements as "Nonviolence!" and "All living beings have same right to live and blossom!" (Næss, 2005, vol. 10, p. 535) articulate norms. But this form of imperative statement is so imprecise that everyone who believes in a slogan could understand it differently. His study of preciseness first emerged from analyzing the vagueness of norms when they are articulated as political slogans (Howe, 2010, p. 378). As a community of philosophers would explain such normative propositions, the explications would increase in length. While the philosopher would consider the statement much improved by the precision of formal analysis, the analysis would unintentionally remove an essential element of its meaning: its political effectiveness. Næss himself knew slogans were imprecise, and regretted having to use them in philosophical works. But he traded precision for political effectiveness upon embracing the political mission of his environmental philosophy.

But it can be dangerous to use simple normative propositions, brief exclamatory slogans, because they are ambiguous enough that some of their possible interpretations can diverge so radically from the utterer's political intent to contradict the original purpose of the norm. The Sierra Club, for example, advocates changing society in a more ecologically harmonious direction in a democratic framework. Yet their "Not Man Apart!" slogan can be understood as a strong biocentric norm. Interpreting this as an ecofascist norm is unfortunately easy: we cannot risk the unity of a human community with its home ecosystem in the service of systemic needs for the freedom and desires of one individual who in part constitutes that ecosystem. But this interpretation has high-profile adherents. James Lovelock, inventor of the Gaia Hypothesis in planetary ecology, advocated that democratic processes are inadequate to repair climate change effectively, because only authoritarian political methods can best prevent humans from doing ecological harm for selfish reasons and act to repair existing damage fast enough (Hickman, 2010). One role for philosophers is, according to Næss, using analytical skills to make these ambiguous slogans more precise. That way, the enemies of the environmentalist movement cannot twist their imperatives into hateful and inaccurate interpretations that, because of the vagueness of political slogans, are nonetheless believable.

Næss's second task for the environmental philosopher is translating precise philosophical analyses back into political norms, and expressing them in a manner that prevents ambiguities, which may lead to self-contradiction. Næss believed he had achieved this task with his philosophical analyses and constructions, particularly his personal ecological philosophy, called Ecosophy T. In Næss's thought, an ecosophy is a kind of total view whose primary concern is ecological and whose normative aspects are so well understood as to be "directly ready for action" (Næss, 1989, p. 37).[5] His term "total view" describes a systematic set of fundamental philosophical positions, which encompasses all that the view holder deems important in the world (Næss, 1964, pp. 22–23). Félix Guattari theorizes ecosophy similarly, as a worldview functioning at the intersection of ecological relationships, social relationships, and the political and psychological forces that constitute your identity (Dosse, 2010, p. 391). A total view is a philosophical system one has developed through one's own thought, "which you feel at home with" (Næss, 1989, p. 37). Ecosophy T unites the principles and concepts Næss learned from complex philosophical analyses of such subjects as Spinoza's *Ethics,* the works of Gandhi, and experimental philosophy, into a complex total view. This process articulates brief imperatives from careful philosophical analysis that has built an elaborate conceptual system to reference in case of ambiguous interpretation.

The philosopher's dual role has been crafted on the presumption of a firm class division of society based on conceptual analytic skill. All people are capable of professing these norms and understanding how to articulate them precisely in their particular projects of living and thinking. But only the philosopher can analyze and construct the systems of norms, which are then put to work as public beliefs. The philosopher becomes a kind of spiritual and conceptual director of society. He does not rule through legislation and force of arms. The social director of thought, in his conceptual role, operates through controlling the meaning of political and social norms. In his spiritual role, a philosopher assumes a kind of religious leadership, but operating through individual counsel. At moments of popular doubt in the prescribed meanings of social norms, the philosopher intervenes at the individual level to restore their faith (Næss, 1965, pp. 72–74).

Political Power in Conceptions of Existence

The ontology of Næss's environmental philosophy is based on the constitutive relation, "We arrive, not at the things themselves, but at networks or fields of relations in which things participate and from which they cannot be isolated" (Næss, 1989, p. 49). He is careful to state that what exists is not a collection of single things isolated in their individuality. All bodies exist as integrated fields, mutually influencing and dependent on each other for their own identity. "The term 'relational field' refers to the totality of our interrelated experience... Things of the order 'material things' are conceived of as junctions within the field.... I interpret this to mean that *the relations* which define the thing *conceptually* converge at the same junction" (Næss, 1989, p. 55). As the building blocks of a total view, one must clearly connect ontological concepts like the constitutive relation to social and moral norms and principles. From the complex and challenging ontology of the constitutive relation, he crafts an environmentalist sociopolitical program with the conviction that the challenge of enormous industry can be met only with the consensus of all individual environmental philosophers and activists. Indeed, he is more interested in the political or social norms that the constitutive relation can imply than exploring its ontological implications. Næss thinks such a consensus is possible around his philosophy, but only because he simplifies its complex concepts into a form that nonphilosophers can easily understand: his normative order, "All living beings have same right to live and blossom!"

This norm includes two concepts, natural rights and blossoming, which need unpacking. Rights can make sense as a privilege to be bestowed or dispensed. States make laws that delineate what cannot be done to those bodies to which those laws apply, and the claims that can be made when those bodies are violated in the proscribed ways. In this conception of rights, the power to bestow rights on some group of bodies rests entirely in the hands of state legal institutions and the people who hold power there. The institution produces no motivation beyond the desires of the already empowered to extend legal protections to some group that is currently not enfranchised. And one could always prefer not to. Conceiving of rights as a privilege to be bestowed is simply not as effective as how the language of modern political activism builds the concept. The activist conception of rights treats them as properties that all bodies of a particular kind possess primitively, analogous to their color or size. Under the privilege conception of rights, a political movement is about changing the desires of the powerful. The activist conception is about convincing people to recognize some preexistent, but long-ignored, property of some group of bodies. Næss believes that the only way to motivate effective popular politics is through activists' belief in rights or value as a naturally occurring property of bodies themselves (Næss, 2005, vol. 10, pp. 163–164). This conception of rights is a legal articulation of the property conception of a body's intrinsic value, and therefore suffers from the same problem: rights are not primitive properties that exist independently of our recognizing them.

Beyond this problem, enshrining and enforcing a right to live and blossom for all individual creatures and communities of creatures faces important practical obstacles. Blossoming is a creature or community unfolding its potential capacities to the fullest extent physically possible. This sounds fine on paper, but political and moral norms that universally respect blossoming often result practically not in an ideal harmony, but great misery. Erazim Kohák describes a public outcry in the 1970s against American National Park policies to cull the herds of wild horses in the Grand Canyon. When the cull stopped and the horses were allowed to blossom free of constraint, their population quickly grew to overtake available food supply, and many horses starved to death, for Kohák a more miserable end than a few hunters' bullets (Kohák, 1984, p. 98).[6] Human hunters were not disruptive interlopers alien to natural processes, but functioned within the Grand Canyon's ecosystem as the horses' predators, our scientifically planned culls constraining their numbers within the limits of their food supplies through the relative mercy of some individuals' quick deaths by the gun. Næss understands such limitations on his norm of blossoming, and addresses them with the same kind of qualification he gives to every element of his ecosophy when considered in the abstract. The norm to respect blossoming "is not some kind of unconditional *isolatable* norm to treat everything the same way. It is only a fragment of a total view" (Næss, 1989, p. 167). A total view unites concepts in a manner that may leave a philosopher scratching his head at the paradoxes that emerge from the compromises of reasonable practice.

In an article about how wolf populations and sheep farmers in rural Norway share territory, Næss describes how sociopolitical policies can handle the fact that practical action cannot permit following the principle of blossoming to its ideal limit. The image of nature as a harmony disrupted by human industrial activity does provide Næss a guiding notion. But his solution to the problem of how farmers, sheep, and wolves share their territory does not aim to restore some ideal condition prior to human arrival in Norway. Instead, we make a political calculation that considers equally the interests of each relevant community. The solution is a compromise, with wolves allowed periodically to feed upon sheep, farmers accepting lower levels of profit, and sheep having to live with occasionally dying to feed wolves and humans while under protection from undue suffering (Næss and Mysterud, 1987, pp. 27–32). On this compromise, the wolves are allowed more leeway to eat sheep without disruption as a recompense for a dangerously low wolf population. At the time of his writing, there were only 5–10 wolves in all of Norway (Næss and Mysterud, 1987, p. 22).

Although Næss sometimes romantically describes wild nature as an Edenic harmony, his actual political prescriptions understand that such harmony is only possible among creatures whose activities never conflict: pure naiveté (Christ and Rountree, 2006, pp. 187–189). There is a broad consensus in environmental philosophy that treating the world only as a resource for humanity's projects is immoral (Plumwood, 2000, pp. 296–297). But organisms living in their natural ecosystems treat other creatures as resources. Food is an obvious example, because aside from those microbes that feed only on nonliving molecules like sucrose, and plants generating energy from photosynthesis and nitrogen absorption, all organisms eat other organisms to survive (Thompson, 2007, p. 158). The relationships among organisms to sustain their lives are not always matters of fatal consumption. Vegetarian sheep farmers will still use their wool for clothing. Grass is a resource for the sheep, as is the human institution of the family farm that protects them from suffering injury by wolves (Næss and Mysterud, 1987, p. 27).

So the result of a politics based on a universal norm that all must live and blossom with equal opportunity is that no community or individual can allow itself to blossom to its fullest extent. Instead, all populations continually negotiate a variety of compromise positions among their conflicting interests. At first glance, the result of the principle appears to contradict the principle itself. One might think that a philosopher could not hold to a principle that all bodies have an absolute right to live and blossom while also endorsing this ecosystemic politics of compromise. Each inevitable sacrifice accompanying any compromise would be an absolute sacrifice. If one must preserve what is valuable, then all bodies having value must imply that all bodies must be preserved. But acting to preserve all bodies results in their destruction, because most forms of eating would be immoral.

Compromise among all cohabitants is the only practical way to enact such a doctrine.

The compromises of worldly action reconcile the absolute and practical conceptions of singularity. Practical difference is evident from the collision of bodies in the world, such as the sheep farmers living in the same territory as wolves, and protecting their sheep with decimating force. The interplay of practical differences constitutes all the beneficial and destructive relationships among all the bodies in the universe. Recognizing that all bodies have a value intrinsic to their existence introduces a normative imperative to preserve that value into our ontology. That which is, a body's value, determines what we ought to do. We obey the norm's imperative by preserving those valuable bodies to the greatest degree possible. That requires compromises like what Næss describes. It is impossible to preserve literally all bodies because organisms must consume others to preserve themselves, and no body can be preserved forever. As well, preserving bodies in their current state would prevent the production of new bodies, which would be just as absolutely valuable as the old. But an ecocentric morality motivated by a belief in intrinsic value can accomplish the greatest possible preservation. This philosophical vision of politics as pragmatic compromises to maximize preservation of the valuable unites the absolute and practical conceptions of the value of singularity. One believes in the intrinsic value of all singularities, but having accepted that absolute preservation is impossible, one works to preserve all that can be, while encouraging the production of new singularities. One manages process for the sake of diversity, the variety of singularities.

However, while Næss's vision of the philosopher as ecocentric spiritual and conceptual advisor will build a class that transparently and democratically steers a society's thinking, or which institutes an insidious form of secret totalitarianism. An idea is politically most powerful when questioning the idea would be as strange as questioning breath. This is not about building a cadre of dedicated followers who accept a political program's dogma as indubitable truth. This vision shores up the contention that environmentalism is ecofascism, which suppresses individual freedoms of thought, speech, and action for the sake of ecosystemic health. The enforcement of a single dogma on a human population ends democratic deliberation and any politics of genuine consensus (Ferry, 1995, pp. 82–83). Fascistic social mobilization is ultimately self-defeating, generating the internal corruption, and external resistance that destroy the goals for which mobilization begins. Like any sensible political activist, Næss understands this. Even in the radical transformation of human society along ecocentric moral principles, liberty must be preserved (Næss, 2005, vol. 10, pp. 570–571).

One does not need force of arms to enforce a dogma. Enormous industry creates images and events of such theatrical terror and absurdity that it impresses a feeling of intense urgency on an environmentalist philosopher. Activism strives to match the accelerating pace of industry (Næss, 2005, vol. 10, p. 194). For human civilization to become ecologically sustainable as fast as possible, a universal ecocentric morality must be adopted quickly. However, it must not be so quick, as in a fascist mobilization, that the movement itself breeds popular resistance. An entire society of people trained and persuaded through formal education, media manipulation, and the daily influence of one's family, friends, and peers who have already been convinced, will live with biocentric or ecocentric moral attitudes at the centre of their lives. The environmental philosopher as ecocentric spiritual and conceptual adviser would be the designer of this regime of influencing popular thought. Only in acute cases of rebellion would the environmental philosopher intervene in an individual case, as Næss described in his analysis of Gandhian face-to-face political action. This is a social activism free of violence and coercion (Næss, 1965, pp. 72–73). But it is not a democratic social activism because the wider population must not know that their morality is the product of manipulation, or they would rebel against it, even if rebellion goes against their physical best interests.

We can unite respect for democracy with ecocentric moralities, although neither with the speed of fascist mobilization nor the comprehensive effectiveness of institutional thought control. Society changes through the inspiration of living exemplars of an ecocentric morality. Live yourself according to the ontological principle that all things have intrinsic value and the moral norm that all living bodies have a right to blossom; the charisma of your living demonstration will encourage the lifestyle to catch on in society. Each exemplar would live a slightly different ecocentric morality, unique to her own personality and the demands of her own life situations. Those who follow an ecocentric exemplar will likewise vary their morality from their chosen model.

Næss himself anticipated this in how he described Ecosophy T, deep ecology, and philosophy itself. He did not want people to follow it directly, but to create their own ecosophy for their own situations, having been inspired by his model. Each person's philosophical worldview need not be consistent with every other in all individual elements of doctrine; they need only carry out compatible practical activities (Næss, 2005, vol. 10, p. 105).[7] Exemplary behavior is the living articulation of metaphysics through one's own identity. Insofar as environmentalist political activism focuses on constituting new identities for people through inspirational behavior, its moral thinking, and political activity have become a matter of ethical philosophy. Persuading people to change their lives through philosophical argument provokes counterargument and backlash, slowing down the process of social change that the environmentalist believes to be so urgent. The singular personalities of ecologically mindful individuals leading exemplary lives become models for those who have yet to give environmental matters much thought (Andersen, 2010, p. 44). The inspiration of living exemplars is the only way society can change while respecting freedom of thought.

Even so, the method of exemplary inspiration as Næss understands it remains remarkably classist. Society is divided into the exemplary ones and the followers, those who absorb the ideas that the lives of the environmentalist exemplars demonstrate, and make them unquestioned dogmas, that over which it seems senseless to argue (van Boeckel, 1997). Philosophers bridge the gap between environmentalist exemplars and followers, while themselves belonging to both camps. They are able to understand the complex and precise principles behind the simple, slogan-like norms such as "All living beings have same right to live and blossom!" The exemplary philosophers would have the mystical experiences of intuiting wonder at the harmonious goodness of nature, then formulate norms based on these intuitions. Follower philosophers would build systems of argumentation to justify those norms, make the norms more precise, or apply those meanings to more particular or situational problems. The role of the nonphilosophical populace would be to accept these moral doctrines without question and live according to their dictates. Here the skeptic must speak.

Intuitions vary considerably across and within cultures, so a single person's intuition has no better chance of insight into genuinely universal truth than some other person's contrary intuition (Nichols et al., 2001, pp. 429–460). What seems intuitively obvious for one person might seem absurd to another. Intuitions cannot conflict if people are to carry out a sociopolitical program without doubt. The follower of a cause must believe that her cause is true and all others false or inadequate. Each follower must understand the cause in the same way. Large numbers of people must transform their thinking at the rapid pace that the environmentalist mood of urgency mandates. Yet an intuition is an experience that calls to mind what you find obvious, what you have never conceived as being sensible to call into question: your dogmas (Weinberg, 2007, pp. 337–338). So the environmentalist goal of fostering in people intuitions of the beauty, harmony, wonder, and goodness of nature is a social program of changing that which an ordinary person finds obvious. There need be no reference to any actual facts, only the institutional and philosophical machinery to convince people of dogmatic principles. One does not need reference to facts to convince someone that an idea is true.

But none of these elaborate machinations of social engineering are actually required, when the destruction that enormous industry causes can be measured practically. Having done that, we can experiment with new ecologically sustainable ways of life. But preventing enormous industry's large-scale destruction does not require such an extreme transformation of human thought as in the vision of the philosopher and environmentalist exemplar as social engineers. We need to only perceive the harms done, understand what causes those harms, and experiment to end them, while maintaining and spreading what benefits of industry we can produce sustainably. The harms of enormous industry are not just those obvious examples that strike us with their immediate theatricality, like the ships anchored in an Uzbek desert, and the oil slicks in the Gulf of Mexico so large as to be visible from space. Ecological science teaches sensitivity to subtle relations and interdependencies, which can accustom people to look for more insidious harms at equally large, if hidden, scales. In this regard, philosophy can serve a practical purpose, in which the ontological concept of the constitutive relation is vital. Constructing a metaphysically systematic philosophy is one way to integrate ontological and moral concepts into an adaptive program for both theoretical and practical thinking. A program with such a purpose is the construction of a subjectivity. So adaptive political programs are not applying moral rules and norms, but putting ethical philosophy into practice.

Practice also provides a defense for the concept of total views or philosophical worldviews in general. It is poor practice in academic philosophy today to speak of having a worldview; such words betray amateurishness. Philosophy today consists of subdisciplines whose boundaries are shaped by conceptual and scientific inquiries. Contemporary philosophers investigate problems; they do not construct worldviews. Only people outside professional philosophy speak of developing a worldview. Yet these are precisely the people who, at this moment in the historical development of the discipline of philosophy, are in a position to judge it. Remember that the UNLV philosophy department was almost destroyed on grounds that the institutional practice of philosophy was no longer relevant to the material practice of public life.

Næss developed his concept of the philosophical total view or comprehensive worldview as a tool for professional philosophers to inform the practice of ordinary people.[8] A subdiscipline of philosophy cannot be part of a political movement if its message to nonprofessional philosophers is that their less rigorous way of thinking about ideas is not welcome in the philosophical community. If the concerns and habits of ordinary people are of no relevance to philosophy, then philosophy is irrelevant to ordinary people. If environmental philosophers want to take their heritage as a political movement seriously, then its ideas, even applied in amateurish ways, should be able to enter the daily habits of ordinary people's thinking. The UNT philosophy department's transdisciplinary engagement with scientific research is one illustration of philosophy's relevance to people's material lives and concerns. Another illustration beyond the university institution is developing environmentalist exemplars, which inspire other people to transform their daily practice in a more ecologically mindful direction.

The historical analyses of environmental philosophy have identified those elements of the Western philosophical tradition that have helped produce the cultural blindnesses responsible for the contemporary ecological crisis. It takes more effort to understand the convergences and similarities of humans and nonhumans, as opposed to their radical difference. Humanity is different from that which is not human, but this difference must not be understood in a way that alienates humanity from the nonhuman, or (the even greater mistake) collapses the nonhuman into a single category and ignores the diversity of existence. Philosophy's role in the practical environmentalist project is to develop a way of thinking that is open to the world's multiplicity of difference and divergence, understands humanity as part of this world, and appreciates the interdependence and integration of all the elements of existence. I believe a philosophy based on the concept of the constitutive relation can achieve this. My goal for the rest of this work is to build such a philosophy, which is itself a new vision of what humanity can be.

Chapter 4

The Essence of an Ecological Philosophy

Living with ecological consciousness requires a moral change, but fundamental changes in morality will not take root in us unless we fundamentally change our entire self-conception. The rest of the book will map this profound ethical task. The foundational step is building a framework to understand reality in terms of interdependence and mutual reliance. Ecology is unusual among the sciences in that it offers a concrete and unambiguous lesson about the nature of reality. Ecology is the science of relationships among dynamisms at the scale of human existence, relationships that demonstrate the interdependence of bodies in the world. The primary principles of ecology are that no body can exist in isolation because its generation and continued existence depends on its relations with bodies that share its history, and that those relations constitute its singular identity (Næss, 1989, p. 49). This scientific lesson is also a moral lesson in that it informs the practice of daily life. The interdependence of bodies for their own and others' existences teaches that all concerns are integrated. Consider the relationships of predator and prey, the most colloquially common example to illustrate ecological dynamics. Even though the members of one group kill and eat the members of another to survive, each population depends on the other to thrive: catastrophic collapse of prey populations result in starvation of predators, and catastrophic collapse of predator populations lead to prey populations overconsuming their resources and starving later. More directly beneficial relationships, such as that of a human and her gut flora ecosystems, are more striking examples of interdependence.

But looking for beneficial relationships alone blinds us to the depth of interdependence even within parasitic relationships. What happens to one element of a system affects what happens to all others. Because of this mutual dynamic affectivity among all constituents of a territory, all these elements integrate with each other. Ecosystems are the product of billions of overlapping feedback loops: bodies constitute their environments while their environments likewise constitute them. This does not unify ecosystems into superorganisms, as in the earliest ecological theories. Interdependence implies coconstitution of the individual with her ecosystem, society, and planet, and ultimately that relation is the fundamental constituent of all bodies.

To Sketch an Ecological Subject

The rest of this book uses concepts from philosophy, biology, ecology, and the related sciences of complexity to plot a path from the ontological principle that interdependence implies coconstitution to the ethical conclusion of an ecological conception of the self. Such a self-image is the foundation of incorporating the moral reasoning to live an ecologically sustainable lifestyle into your daily existence. Immanuel Kant was right to believe that you can reason morally while holding morality in contempt. But someone like that will only follow through his reasoning with action, at best, begrudgingly. For morality to have any significance for you as an interest in your life, you must consider yourself a moral person already. Ecocentric moral reasoning faces the same problem. Your interdependence and integration in an ecosystem must inform your sense of who you are for ecocentric moral reasoning to matter for your day-to-day existence. Ethics, the philosophy of what kind of person you are, is the ground of moral practice. We cannot change our character as a social being without changing our individual self-images. A project to change your self-image must analyze any concept that may contribute to how you think of yourself, such as the nature of the universe, your place in it, what we know of the world, how we know it, and how we come to be. All movements for social change depend on this ethical event at the individual level, examining and changing yourself to achieve the practical goal of making human civilization ecologically sustainable.

According to one standard view in moral philosophy, only a discrete individual who is sufficiently autonomous can make self-conscious decisions and accept responsibility for them. This concept of the discrete autonomous individual is an *egoistic* concept of subjectivity. But an ecological philosophy conflicts with egoistic self-understanding. The concept is useful for some political and moral problems, but our philosophical priority must subordinate egoism to an ecological conception of subjectivity. A long-standing principle in Western philosophy is that you cannot derive an ought from an is: premises about what is can lead to no valid conclusions about what is morally right or wrong. But strictly separating philosophical domains in all cases restricts our inquiries from understanding problems in the world, which require using diverse concepts. One of these is humanity's ecological crisis and ecocentric moral thinking. Val Plumwood writes:

> Mainstream environmental philosophy is problematic not just because of restriction *in* ethics but also of restriction *to* ethics. Most mainstream philosophers continue to view environmental philosophy as primarily concerned with an extension of existing ethical frameworks. For example, instrumentalism is viewed as a problem in ethics, and its solution seen as setting up a theory of intrinsic value. But this neglects the key further aspects we have been examining, of dualism and the account of the self and of human identity as hyperseparated from nature, the connection between this and the instrumental view of nature, as well as the broader historical and political aspects of the critique of dualism and instrumentalism. (Plumwood, 1993, p. 173)

Plumwood's hyperseparation is the absolute dichotomy of human and nature into two categories with infinite conceptual distance between them. Her point and mine is that understanding and solving the major moral and political problems of environmental philosophy requires employing ontological as well as strictly moral or ethical concepts, understanding how the use of one domain's terms affects how you use the other, and accepting that the absolute dichotomy of humanity and nature stands in the way of an ecocentric vision of humanity and human society.

Morality's Hidden Ontology: Overcoming the Dualism of Active and Passive

The absolute dichotomy of humanity and nature encourages us to map other dualist false images onto its frame. One dry dualism that frequently occurs in humanity's environmental thinking since the ancient Greeks associates activity with the human and passivity with the natural. Humans are the only creatures that act, and all other bodies react and are acted upon (Plumwood, 1993, pp. 107–110). An ecological philosophy's focus on interdependence and the coconstitution of all bodies and systems, is a stark alternative to that dry old dualism. All that matters in dichotomous thinking is the essential meanings of the two categories, and the proper slot for each body. So vast diversity in singular identities will be ignored, subsumed into their proper side of the dichotomy. What matters for an ecological ontology, in contrast, are the actual relations that assemble the parts of systems. Instead of mapping the world through the lens of a simple, abstract concept, ecological thinking investigates each body and its constituent relations in their singular existence, and derives general framework concepts for understanding existence from trends and tendencies among many bodies. You, a subjectivity, are a complex assemblage of a plurality of forces, which develops a sense of selfhood.

Understanding bodies as processes means everyday words like "part" and "whole" will have very different meanings. Every body is a whole, constituted from parts, and each of those parts is a whole with its own parts, and so on. At no point can you discover a body that is truly simple. Such a unity having no component parts is impossible. Identity is not a simple unity, but a fluctuating set of "networks or fields of relations in which things participate and from which they cannot be isolated" (Næss, 1989, p. 49). Physical processes assemble the parts of a body that constitute its properties and identity. But one body can be part of another without being physically inside it. Interdependence means that each body in a relationship depends mutually on each other for existence; changes in one body causes affects that cascade through their interdependence relations. So interdependence relationships are constitutive pathways for affects that create and change bodies and relations. Nothing can ever be completely autonomous in the strongest sense.

If X and Y are related somehow, then a change in X will change Y. The relation between them constitutes, at least in part, both their identities. So while X and Y are separate bodies that you can legitimately regard as discrete entities, their identities are integrated insofar as a change in one affects change in another. If X changes, the resulting change in Y need not be anything like how X altered. That there is change is evidence of their mutual affectivity, that they coconstitute each other moment to moment. This principle of the mutual affectivity of all change underlies the environmentalist moral injunction to act mindfully of how your actions change surrounding bodies. We can defend it with a simple recourse to practical reason, avoiding the destruction that comes from acting without heed to possible consequences. The changes you make in the world can turn back onto you unpredictably, so watch out! This practical reason often carries the most political effectiveness precisely because it is easily sloganized. But the ontological concept of the constitutive relation transforms this ordinary injunction into a profound worldview. A prudent person becomes an ecological subject.

A maxim of simple prudence is easy to follow, but also easy to ignore if nothing else of how you understand yourself changes. Understanding yourself as an inherently ecological body is a difficult revolution in your self-image, but it affects your deepest motivations. The typical subjects of ethical thinking and moral systems are human persons, and persons' natures and capacities determine moral duties and obligations. Speaking within that typical context alone, the human person is the indivisible atom of a moral calculus. The person holds rights, follows duties, joins contracts, and embodies virtues. The person desires, wishes, and thinks. The person is the elementary particle that constitutes families, communities, states, and societies. But thinking only of personality, politics, and society does not offer the most complete understanding of everything that constitutes a human person. These psychological and social domains are the products of processes that exist only at the forefront of a human's self-consciousness. The conception of the self as an ecological subject goes beyond the possible reach of moral injunctions to understand humans not only in our social and psychological dimensions, but also our ontological aspects as aggregations of physical forces, bodies, and properties.

Some examples of a person's ontological constituents: the construction of membrane-bounded cells from an amorphous chemical soup, the intricate folding and accurate reproduction of chromosomal DNA and proteins, the formation and functioning of organs, the folding of cellular walls to create blood vessels, tendons, muscle, ligaments, bones, and the neuronal structure of the brain and perceptual system (DeLanda, 2002, pp. 62–64). These processes are themselves constituted from even smaller scale processes of molecular and atomic bonding, which produce cells and the bodies within cells. A human body's internal processes do not constitute only itself. Colonies of microbes live inside the human digestive system, facilitating the nutrient absorption process in a symbiotic relationship that allows the microbes and their human home to thrive, an ecosystem within each of us (Cockell, 2005, pp. 375–290). External processes constitute a human body as well. A human exists within an ecosystem, and came to be from a collision of a variety of bodies in that ecosystem: the chance events that resulted in conception, the food a mother digests that nourishes her embryo, matter being cycled through her body to grow a fetus.

These ecosystemic processes are ultimately cosmological in scope. The constitution of a human body is one relatively brief phase of the ongoing fluctuations of matter in an ecosystem (Leopold, 1970, pp. 111–115). That ecosystem, in which a person is a small part, developed in the dynamic and often unstable processes constitutive of the planet Earth (Ward, 2009, pp. 81–89). Earth and its solar system are just one small part of the processes of stellar formation and decay constitutive of almost the entire history of the universe. Energy generated from stars thousands of times more luminous than the sun and their supernovae constitute interstellar dust clouds stretching across a galaxy. The movements and interactions of these dust formations and the stellar energy fields that animate them are what Lee Smolin calls the ecology of a galaxy (Smolin, 1997, pp. 120–122). Earth was assembled from these clouds, and its ecosystems and organisms are the minute details in the history of this ball of condensed dust.

The processes from which a human body is assembled flow back and forth inside and outside that body. People around each other breathe air that joins with their very blood before being exhaled (Kretz, 2009, pp. 124–125). Your own self-image is constituted through yet another order of community processes: social interaction within a family, a community, a state, a global aggregate of people in varying intensities of solidarity, linked through multiple platforms of communication media. These external relations with social processes and institutional structures constitute your self-conscious identity (DeLanda, 2006, pp. 56–57). Moral philosophy operates in social and political contexts, while ethical philosophy operates and revolves around understanding yourself as a person. So moral and ethical philosophy typically conceives of a person as a spontaneous source of action with an important degree of autonomy from forces that would otherwise constrain her into passivity. Processes across many domains from the chemical to the cosmological constitute the physical existence of an individual human, yet are not usually part of self-conscious decision making.[1] Such processes change in response to a person's environment, but there is typically no deliberation or self-consciousness involved in such change. As such, we usually consider those processes passive, immaterial to making morally and ethically valid actions and decisions. Yet without these processes, there could be no bodies capable of self-conscious decisions.

The Relation, the Atom of Everything

Most moral thinking and deliberation requires considering the person as an autonomous unified body. While perfectly legitimate in that context alone, this is not a complete vision of a person because it ignores all the other relations that constitute a person as a relatively stable body in the ecology of a fluctuating universe. A naive unity is a conception of any body as autonomous and unified without giving any thought to the physical complexity of its structure or its historical processes of generation. Such a conception is naive because it does not consider a body in terms of its relationships with the processes that generate it, and the processes in which its activity and existence is always integrated anyway. Processes generate all bodies, humans included. These processes are not separate from the body they generate, but are themselves parts of it as they maintain the relations among themselves that constitute the whole. A body is never truly complete or finished; stability and good health is the product of ongoing maintenance processes. A body is not built to completion according to a predetermined plan, but a continuing process that eventually fails and disintegrates. Death is no culmination, but breakdown.

Although organic bodies are most obviously processual in character, all bodies are constituted as processes. The designation of one body as part of another therefore depends on whether it plays a role in the other's generation or maintenance processes. Processual part-whole relations pay no heed to whether a whole physically encloses its parts. A process that helps constitute a body may remain entirely external to its physical boundary, perhaps instead providing a context that constrains and opens possible directions of its development (Guattari, 1995, pp. 34–35). Flows of matter and energy that are constitutive of an ecosystem are one example of such an external system actively developing a body within it, such as a single organism making a home for itself there. The larger flows of an ecosystem are part of a single organism that lives there. The development of its constitutive processes physically integrates a body with all processes surrounding it and within it.

The universe is inevitably and immensely complex. So conceiving of something as a naive unity is useful for bracketing purposes, to focus on a problem immediately to hand, precisely because such restriction of focus passes over the actual complexity of a situation. Selectivity is essential to the pragmatism of ordinary perception: we always focus on particular aspects of our surroundings so that we can act without being overwhelmed by a barrage of information not directly related to the tasks at hand (Wolfe, 2010, pp. 112–113). Such pragmatism of selectivity is not only a matter of everyday life. Consider the following abstract structure as a useful illustration of how conceiving of a body as a naive unity can be productive in scientific knowledge, even in its supposed drive for total comprehensiveness.

If I want to study the constitution of a body J, I can consider it an assemblage of its constituent parts K_{1-n}. J would be the complex body, and to study its structure I can consider each of its parts K as naive unities, because the complex structure of K does not matter to me right now, only that of J. If I want to consider the complex structure of one of the bodies K, I can conceive of each K as constituted from parts L_{1-n}. Likewise, considering the complex structure of a body L requires me to take its constituents M_{1-n} as naive unities, and so on. This account diagrams reality like the nested wholes of Matryoshka dolls. But the image of nested wholes does not capture the complexity of the relations among the various microscopic, mesoscopic, and macroscopic levels at which we can analyze bodies. A whole can also have parts that overlap with other wholes, or are tangential to the whole in question. The processes that generate J need not only occur at K. Processes in L that generate K can also have a direct or indirect role in constituting J, could also help constitute J. The same applies to processes in M, N, O, P, or levels that are macroscopic relative to J, like I, H, G, and so on at any other Matryoshka level I could mention in an abstract structure.

Our institutions of knowledge in the sciences and the humanities must reflect the complexity of the phenomena under study. The disciplinary boundaries of professionalized knowledge carve the universe into distinct domains, each field having its own methods, tools, and practices. Yet affects and the processes that constitute them proliferate without heed to disciplinary boundaries, or distinction of levels of analysis. This does not imply that the divisions of the sciences are somehow illegitimate or a fiction. The effectiveness of the disciplinary sciences clearly refutes any reductive or eliminative account of the social construction of science. After all, just because social processes constitute some complex body like an institution or a scientific discipline does not mean that it is not real, or less real than bodies made from different processes (Hacking, 1999, pp. 64–66). Our powers to understand the universe depend on the intellectual efficiencies of the disciplinary system. But tracing affects among relations of complex systems is inherently transdisciplinary.

Manuel DeLanda describes the universe as "a contingent accumulation of layers or strata that may differ in complexity but that coexist and interact with each other in no particular order: a biological entity may interact with a subatomic one, as when neurons manipulate concentrations of metallic ions, or a psychological entity interact with a chemical one, as when subjective experience is modified by a drug" (DeLanda, 2011, p. 6). That real processes are transdisciplinary is also the central principle of Félix Guattari's approach in *The Three Ecologies*, where he examines the integration of psychological, social, and ecological domains. The constitution of an individual personality "establishes itself at the crossroads of multiple components, each relatively autonomous in relation to the other" (Guattari, 2008, p. 25). The activity of chloroplasts in one of the leaves of a bush helps constitute a key process in maintaining the whole of which it is part. Understanding relations of parts and wholes as a Matryoshka, the chloroplast is part of a cell, the cell is part of a fiber, the fiber is part of a leaf, the leaf is part of a branch, and the branch is part of the bush. But because the activity of the chloroplast maintains the bush as a stable structure, the chloroplast may be considered part of the whole bush. And that chloroplast is also part of the whole forest glen where the bush grows.

Mereology, meditation on relations of parts and wholes, is one of the most abstract, obtuse, and difficult topics in ontological philosophy. It was a centerpiece of the ontology of Aristotle, and has been a perennial subject for philosophy ever since. The particular mereology at the center of my analysis is peculiar in its focus on generative processes, and has several important implications. One is that relations that arise in the generation of bodies introduce temporality into the part-whole dynamic. Speaking in process contexts, a body is part of some other body when it plays a role in the other body's generative processes; so a part and the whole it constitutes may not exist simultaneously. For example, a body's activity may generate a process that constitutes some new body, but is itself destroyed before this new body emerges. Nonetheless, because that initial activity was a condition or cause of the new body's generation, a processual mereology makes that activity part of the new body. Generative mereology also implies that one body can be part of another while being external to it, or even remarkably far away. To illustrate, consider how each galaxy's stability depends on a gravitational field that extends throughout it, which is constituted by each body in the galaxy. So the gravity of a single star on one side of the galaxy directly contributes to the field that conditions the generation of bodies thousands of light years away from each other. So proximity or containment is not required to play a role in a body's generation either.

Perhaps the most truly radical and characteristically ecological implication of thinking about existence as the generation and collision of processes is a universal holism that contains genuine multiplicity. Because physical containment and proximity in space or time do not limit whether processes can interact, we can ultimately consider all processes and bodies to constitute the entirety of the universe itself as a unified whole. Every process ongoing today in the universe was generated by prior processes, making these earlier bodies part of contemporary generative processes, all the way back to the beginning of the universe. All processes that exist today in the observable universe (at least) have their origin in the Big Bang, the generation of everything. The cosmic background radiation that permeates the entire universe is an event from that origin interacting with all bodies in the observable universe. These processes and bodies alone are sufficient to consider the entire universe, whether in its complete history or at any duration or snapshot moment of its existence, an integrated whole.

I should elaborate another implication of generative mereology, that every body is a complex whole at every level of analysis, and so there are no primitively simple bodies. As an example, consider once more a human body. No component of a human body is its absolute ontological foundation, whether in its physical makeup or its self-conscious personality. Such an absolute ontological foundation would be a naive unity, simple body that is not constituted by some other process or processes. All of a body's properties are constituted through internal or external relations with other processes. From this conclusion, we must accept that the fundamental condition of all being is the relation, which constitutes bodies and processes.

Combating a Counter-Productive Conception of Properties

Properties are assemblages: either complex bodies or processes of interaction among bodies. This irreducible complexity implies that simple universal words, such as "green" or "round," which are often used to describe properties, do not refer to bodies as simple as the words. Universal property terms are a convenient shorthand for practical descriptions, because the most accurate descriptions of phenomena and bodies are potentially endless. This conclusion stands against common practices of thinking about the fundamental nature of bodies and their properties in both analytic and phenomenological philosophy.

Because a body is the sum of its properties and their relations, an exhaustive list of all a body's properties completely describes its identity, defining precisely its singularity. Yet unlike what our language suggests, a property is not a simple unit. Ontological philosophy that follows the conventions of our language builds a massive disciplinary edifice on this mistake. For example, to say that an apple is green concisely describes the property, so analytic philosophy's common parlance considers the universal term "green" one of the apple's constituents. "The strongest version of the bundle theory of substance claims that of necessity, the substances that make up the world are bundles of universals" (O'Leary-Hawthorne, 1995, p. 191). The universal would be more a more fundamental constituent than any physical component like its seeds and skin. These universals would exist in a one-to-one relationship to the property words of language. A universal is a simple entity that would bind together with other such simple entities as *round*, and *sour* to constitute the apple in question.

But a universal is not a constituent of an apple. The color of the apple is its property, but a property is not the primitive entity, some universal in which the bundle theorist believes. Color is generated by a continuing physical process that articulates itself through the relation of the colored body, an observer and her perceptual apparatus, and the electromagnetic fields her apparatus detects. The chemical composition of an apple's skin interacts with a field of ambient electromagnetic radiation, a field of photons, in a manner that absorbs all photons except those of a particular set of wavelengths. If a human observer were present in pedestrian proximity to the apple and was interacting with the ambient electromagnetic field, she would perceive the light that the apple reflects, and she would call that light "green" in English. The property of being green is constituted from the relation of all the molecules linked together to form the apple's skin, with an ambient field of light, and an organism capable of perceiving light that has a frequency of about 540 terahertz.

The apple's properties are constituted from such complex relations that it appears improper to say that the apple itself has properties. Where are a body's properties, if not in that body itself? The answer is not so simple, because a body is not so simple. A complex body constituted from processes great in number and variety will generate more processes from this initial convergence. When processes converge, the resulting new processes will likely be even more complex: there will be more processes, and they will have greater variety. Some processes, however, begin to malfunction when they converge, and the whole system of collaborating processes and bodies dissipates. But even the dissipation of a complex process can generate new processes. Consider the decay of a corpse, in which a human body breaks down, but spurs processes of putrefaction: I will become food for worms, insects, and bacteria; and fertilize plant growth (DeLanda, 2002, pp. 115–117).

When we take the concept of the constitutive relation to its logical limit, we should understand what philosophers have traditionally considered primary qualities like weight and mass to be relational. Consider a roundish chunk of dolerite stone. Considered naively, a stone is a simple body with no internal parts. But dolerite is an aggregate of several different minerals, affecting each other in the compound to form a complex solid lattice of their constituent molecules. The atoms of which these molecules are constituted also contribute to the generation of fields of elementary forces like gravity and electromagnetism. On the scale of individual atoms and molecules, these forces are fantastically small. But when enough molecules aggregate that I can hold the collection in my hand, the forces are noticeable to pedestrian perception. The stone's weight, a so-called primitive property, is a function of the interaction of gravitational fields that it, the Earth, and all other surrounding bodies generate. The stone's mass is the combined mass of the atoms that compose it, a collection of all the internal bodies that integrate to constitute the dolerite stone I hold in my hand.

Everyday perception and the language we typically use to describe it can lead to a naive understanding of properties as constituents of a unified substantial object. Edmund Husserl called these kind of simple encounters "an attentive perceiving, I am turned toward the object . . . I seize upon it as this existent here and now" (Husserl, 1982, p. 70). Conceiving of a body as simply being there understands it as a naive unity. We usually conceive of a body this way when we focus on a matter to which that body is a peripheral concern. As an example, a sociologist of gender does not typically consider hormone chemistry when she investigates working conditions in the fashion retail sector. Hormones are chemically important for the construction of gender, but their complexity does not matter to the social and economic structures that a sociologist investigates. Husserl's initial approach to experience was to subject the simple and primitive to an intense analysis that we normally employ for investigating something complex. It was an effort to think of the primitive nature of bodies. In the most simple encounter, thinking only of how an object stands mute before you, questions of whole and part are not important. Only presence matters. Phenomenology is a philosophical inquiry whose goal is to discover the essence of presence (Bello, 2009, p. 60).

A Husserlian approach seeks complexity in an object's simple presence to perception. It begins from what he calls the natural attitude, apprehending a body as a naive unity. Such an attitude can be sensible, useful, and amenable to many people's intuitions about experience. But it is incompatible with a process philosophy such as ecological thinking, where inquiry begins not with presence, but assembly. Analysis of process, where no body is a simple unity, understands properties as constituted from converging relations. The simple encounter is a productive concept for phenomenological inquiries, but is a stumbling block for an ecological philosophy. Although phenomenology's influence on environmental moral philosophy may be infrequent, both contain similar interpretations of bodies as naive unities. There is no need to investigate a naive unity's mute presence; you should only appreciate it (Aikin, 2008, p. 27). However, if you consider the unified appearance of a body in a simple encounter to be the fundamental nature of all bodies, you can miss the complexities of a body's internal structure and external relations and influences.

Understanding existence through a lens of complexity supports the notion that properties are a body's primary constituents, though this is nothing like the sense an analytic bundle theorist would have of the problem. Identity answers the question of what it is to be what one is, "the question of essences — in other words... 'What is X?'" (Deleuze, 1994, p. 188). The debates in analytic ontology between a bundle theory and a substratum theory revolve around whether universal properties alone are necessary to constitute a body, or whether those properties must inhere in some kind of substrate that has no properties in itself. A bundle theory would give an account of a body such as, "Take my neighbour Cyrano: he is of a certain age, he has a big nose, he has such and such a height, and so on. And this is all there is to know, and all there is to *be* Cyrano — his properties" (Benovsky, 2008, p. 176). A property on this view is a simple unity, just like the term in language: "big-nosed," "42-years-old," "six-foot-two." This kind of identity question is too narrow, more a question of identification rather than an attempt to understand the singularity of bodies.

An ecological philosophy that conceives of bodies as products of constitutive relations understands properties and identity differently. My question of how what is comes to be as it is regards generation and constitution. Identity as the bundle-or-substratum debate understands it can play an important role in this question, because an account of becoming must include what becomes. But there is a distinction between how a body becomes what it is, and what a body has become. Bundle and substratum theories focus only on the latter, to their detriment. My concern is what they ignore, the process of generation, because at no point does becoming genuinely stop. At no point can I say of some body that it is what it is and will become no more. Analytic philosophy's property ontology understands identity as static, as if its becoming is complete. But that would imply that a body will no longer change once it is what it is. Particular processes in the body may have stopped at the moment we observe. For example, the apple in my fruit bowl will ripen no more as it would if still growing on its tree. But further processes, such as decomposition or digestion and absorption into my own body, may still occur. When a property such as Cyrano's being six-foot-two is understood in terms of its generation, it is laughable to consider a property term, like being six-foot-two, to refer to a simple unity. Over 200 genes, along with the enzymes and proteins they code, contribute to constituting the height of a human (Mayeux, 2005, pp. 1404–1407). Analytic bundle theory is unprepared conceptually to understand the processes that generate the properties of bodies.

All bodies are constituted from other bodies internally, externally, and bridging internal and external. The examples of the human body and a dolerite stone described the processes constituting these bodies as scientific investigations have discovered them. Speaking more generally, a constitutive process is generated when material and energy converge to create a pattern whose structure is stable enough to affect and be affected by other such patterns. The pattern's stability does not imply that it must be static, nor is my understanding of a body restricted to one freeze-frame instant. When I discuss a body, I include its full history from constitution to dissolution. The matter and energy that constitute a complex body arrange themselves into patterns of exchanges from interior to exterior and vice versa. The variety of processes that constitute a body produces its singularity.

How You Understand Existence Has Moral Implications

This is why constitutive relations are integrative instead of holist, preserving instead of subsuming multiplicity in unity. This despite the prevalence of holist images throughout the tradition of environmental philosophy. J. Baird Callicott's early philosophy in defense of the biotic community provides a clear example of a morality where the value of the whole entirely overwrites that of the individual. In such a morality, individuals are replaceable by any other individual that would perform the same functional systemic role in promoting overall stability. Moralities of holism conceive of individuals as cogs in the machine of the ecosystem, and only the whole is morally considerable (Zimmerman, 1995, pp. 227–230). Holist thinking at its most straightforward conceives of the whole as a primitive, subsuming its parts into one unit, no matter how complex the arrangement of their integration may be. Simple holism is untenable as a principle of environmental moral philosophy. In political contexts, morally prioritizing the whole at the expense of its constituent parts is a fascist trajectory, a disregard for individual lives. A holist understanding of interdependence and constitutive integration in an ecological ontology is similarly untenable for *exactly the same reasons*.

Processes organize not only matter to constitute each complex body, but also the fields of force that bodies and environments generate. The same goes for a body's properties, whether primary/perception-independent like mass, or secondary/perception-dependent properties like color. Contemporary particle physics investigations have revealed that mass is actually constituted from the energy of the interaction between particular types of fundamental particles. Therefore, at the most fundamental aspects of matter, mass is a property constituted from relations. Understanding a body as a naive unity may be fine to point it out, distinguishing one from another in the confusion of daily perception. Such thinking may have the benefit of simplicity, but if you want to understand the world, then simplifying your thinking is foolish. Complex matters require complex thought (Guattari, 1995, p. 61). The concept of process is better suited to understanding the history of a body's generation, continued existence, and dissolution; its becoming. Understanding the constitutive relation as the active integration of complex networks of bodies preserves the singularity of bodies in our thought, even as we understand their identity as the product of a plurality of processes. The processes constitutive of a singular body are themselves singular bodies, each with their own identities. So one singularity is constituted from the assembly of many singularities.

Ontological considerations can strengthen an ethical or moral stance, illustrating how inquiries in one domain of philosophy can aid inquiries in others. Showing how actual processes constitute bodies in their singular physical identity lets us focus on how bodies are interdependent not only for their survival, but for their identities as individuals. Understanding the nature of bodies this way lets us approach practical concerns having to do with them more productively than if we held a more traditional ontology of bodies as ultimately reducible to simple unities. The constitutive relation is a concept of how bodies are generated and exist, an ontological concept. It helps create a more nuanced understanding of interdependence and integration of bodies into ecosystems. But we must still understand how to think subjectivity, the place and constitution of the thinker itself, ecologically.

Chapter 5

The Conditions of Selfhood

If all the world is process, then you and I are no exception. The subject is an ecological process at every level of our existence. If there is no place for a subject in your ecological philosophy, then you have again made the mistake of building an absolute dichotomy of humanity and nature. Such a mistake would not be obvious at first; its subtlety makes it all the more dangerous, more difficult to catch. An ecological philosophy that neglects accounting for the development of subjectivity is still the product of a subject's own thinking: mine in writing and yours in reading. A genuinely comprehensive ecological philosophy would account for how ecological processes constitute ecologists. Understanding how humanity develops through natural processes overcomes the dichotomy of humanity and nature. Therefore, an ecological philosophy should have an account of the processes that produce humanity: self-conscious bodies that organize socially.

The most basic form of self-consciousness is constituted from the physical feedback processes of autopoiesis. My account of autopoiesis has three stages of clarification, and one of application. I first introduce the concept's complicated history, specifically how it developed in cognitive science. Second, I discuss how systems theory interprets autopoiesis, and critique its idea that all autopoietic systems are operationally closed in a strong sense, communication between them being impossible. Finally, I describe autopoiesis as a field of perceptual affects, and then deploy it as a scientific foundation for Arne Næss's conception of self as a place. Self-as-place is a key concept in Næss's environmentalist philosophy, but he does not describe it with enough detail in his own work for the concept to be much more than a platitude. I believe that autopoiesis, decoupled from systems theory's strongest interpretation of closure, can combine with Næss's intriguing vision to produce an ecological conception of subjectivity. This would include the minimal conditions for subjectivity in an overall process ontology in which the flux of all bodies are contingent thanks to their generation through dynamic relations.

A History of Autopoiesis in Cognitive Science and Philosophy of Mind

Autopoiesis was developed in vestigial form in the context of a broad program of cybernetic research that began in the 1950s and encompassed hundreds, if not thousands, of scientists and theorists. The most important autopoiesis theorists for my project are Humberto Maturana and Francisco Varela, whose major collaborations happened in the 1970s and early 1980s (Varela, 2009, pp. 64–66).[1] Cybernetics was a diverse community of researchers, several projects coming together in a common inquiry to understand cognition. The cybernetics community saw linguists, computer scientists, psychologists, biologists, and philosophers converge onto a single program (Hayles, 2005, pp. 279–280). Its ultimate goal at the time of cybernetics' genesis was to create an artificial intelligence, a machine that could pass a rudimentary version of the Turing Test. Such a machine would be able to understand questions asked of it and respond in a recognizably intelligent manner. This task was primarily understood as designing a machine that could understand language, defined as the syntactical arrangement of propositions and the semantic meanings of words and expressions (Searle, 2004, pp. 47–50). Although the cybernetics community initially focused on this goal, it eventually developed into a loosely connected interdisciplinary network of people who today identify as cognitive scientists (Clarke and Hansen, 2009, pp. 2–3).

Over the decades since its formation, the cognitive science community has developed in a manner that can be confusing for someone trying to use its concepts for other purposes. The most widespread philosophical influence of its artificial intelligence project was the popular image of the human mind as a computer program, the software of the brain's hardware (Fodor, 2008, pp. 7–8). Having conceived thinking as software for the composition of language, mainstream philosophy of mind understands cognitive activity to be primarily the composition of language. Philosophy of mind and philosophy of language overlap considerably in contemporary North American philosophy. The compositional theory of thought and mind dominates these subdisciplines: philosophers understand perception as the translation of sensory data into propositional statements and representations of an organism's environment in its neural architecture (Churchland, 2002, pp. 80–88).

Standing against the compositional linguistic conception of thinking and mind in cognitive science are the enactive theories of mind. Enactive theories take our perceptual engagement with our surroundings, not language composition, to be the primary cognitive activity in the constitution of the mind.[2] "Cognition is the exercise of skillful know-how in situated and embodied action" (McGann, 2007, p. 466). Perception is making sense of your surroundings in terms of what will increase and what will decrease your power, wisdom being to act such that your power increases. It is an account of perception as the power of an organism to discern practical differences in the world. So perception does not involve propositional or properly linguistic content at all, but is entirely a matter of affectivity, activity that empowers an organic body or diminishes its power (Deleuze, 1988, pp. 49–50). The debate between the compositional and enactive camps in philosophy of mind is over whether cognition is a matter of knowing how to carry out activities in the world, as the enactive theorists propose, or the compositional theory that cognition is about representing and remembering facts about the world as propositions. For a compositional theorist, one cannot know how to act without knowing facts; for an enactive theorist, one cannot know facts without the practical ability to move in the world (Fodor, 2004, pp. 30–31).[3]

If I brought my ecological philosophy's conception of autopoiesis to the partisan debates of philosophy of mind, it would be obvious to all that my sympathies lie with the enactive theorists. But my conclusions differ from Varela's, because I am not interested in the same philosophical problems that he is. Varela and I converge at his and Maturana's initial creation of the concept of autopoiesis to describe a body whose own chemical activity produces itself. Varela elaborated autopoiesis through engaging with the problems of cognitive science and philosophy of mind. It was a philosophical component of a biological investigation into the essential nature of life. If metabolic activity is the essential chemical activity of life, then life itself is the physical process of creating a body that maintains its own integrity (Weber and Varela, 2002, pp. 117–118). They examined what physical structures the simplest metabolic reactions constituted. Metabolic chemical activity produces molecules that bind together to act as a wall or membrane surrounding that reaction. This wall is thermodynamically open, its structure facilitating an exchange of energy from outside to inside. Surrounding molecules, which can function as fuel for the central metabolic activity, enter. Waste molecules from the metabolic activity exit (Thompson, 2007, p. 98). That chemical activity also maintains itself because it produces molecules that reinforce and repair its membrane from the inside, preventing the autopoietic body's disintegration (Maturana and Varela, 1980, p. 98). Autopoietic chemical activity produces an autonomous body, an individual that is itself, at heart, that chemical activity (Maturana and Varela, 1980, p. 87).

Varela developed the concept from its foundational formulation through his dialogue with cognitive science. He considered the proper application of his analysis of autopoietic perception to be an alternative account of mind, contrary to the compositional conception of cognition. Varela developed an account of autopoiesis as sense making, the production of significance in the world through the activity of moving in it. Evan Thompson illustrates what sense making is with the example of a self-propelling bacterium swimming toward an agglomeration of sucrose. The bacterium can perceive the presence of sucrose around it, and because sucrose constitutes fuel for its metabolic activity, it moves toward the region with a higher concentration of sucrose molecules (Thompson, 2007, p. 157). His example, that a bacterium can distinguish sucrose molecules from an otherwise useless background, suggests that even the most rudimentary forms of life perceive.

The bacterium's power to distinguish creates a new property in the objects it perceives: significance. The emergence of significance from perception to constitute the primordial order in an otherwise chaotic world is the ontological centerpiece of the enactive theory of mind (Knyazeva, 2009, p. 168). Because significance, the product of organic activity, produces the foundational order of existence, all order in the universe has a pragmatic ground. In distinguishing sucrose from the surrounding useless chemicals, Varela's bacterium organizes the world according to a principle of what its food is, and more complex organisms produce worldly significance according to a correspondingly complex set of principles. Yet as a centerpiece for a theory of mind, sense-making encounters an anthropocentrism problem because Varela, Maturana, and Varela's protégé Thompson develop their thinking in dialogue with a tradition whose focus has historically been on the nature of the *human* mind.

Significance is the primordial product of autopoiesis, emerging from the tendency of a metabolic chemical reaction to consume fuel to preserve its existence. Following Spinoza, Thompson understands perceptual activity as a function of self-preservation, conatus. The enactive theory of mind is the latest development in a disciplinary discussion that began with cybernetics, the investigation into the possibilities for creating an artificial equal to natural human intelligence. Enactive theorists primarily understand their own philosophical program as an opposition to the stale dogma of the compositional theory of mind, whose disciplinary dominance already established the essential power of mind as the production of language. Its default disciplinary context in philosophy of mind frames conativity as the possibility condition for the complete human range of semantic meaning and propositional syntax (Thompson, 2007, pp. 157–158). Therefore, if conativity alone is the ground for producing the meaning of words and the syntax of linguistic expression, then all organisms can compose some sort of linguistic propositions. The human mind becomes the paradigm for the minds of all organisms (Thompson, 2007, pp. 161–162). Although Varela himself did not endorse such a simplified conclusion from his work, Thompson was not afraid to advocate that all life forms have primitive linguistic capacities (Protevi, 2009, p. 104).

Although some species may have linguistic and cognitive powers that parallel humanity's, our own syntactic and semantic creative powers are unique to our own kind. They are certainly of an entirely different kind than the sense-making perceptual powers of different organic kingdoms on Earth, like fungi. This is an empirical matter. The ground of Thompson's error that all organic cognitive powers are variants of the human linguistic model does not lie in the logical steps of his argument, but their disciplinary foundation in the historical practice of science and philosophy. Enactive theorists of mind study the physical phenomena of organic conativity: autopoiesis, an organism's perceptual activities in its environment, and the structure of perceptual apparatuses. Compositional theorists study the neurological underpinnings of syntax and semantics in a philosophical framework of the computer metaphor of the mind's parallelism of language and thought.

Because both groups developed in the cybernetics community, they both thought they were building competing conceptions of what the mind was. This was the original purpose of cybernetics as a new discipline when the community assembled itself from an interest common across many scientific disciplines in the nature and construction of artificial intelligence, artificial minds. Enactive and compositional theorists of mind see themselves as advancing competing definitions of what the mind is, but any common object called mind disappears as their fields of research diverge. Manuel DeLanda describes the conflict as a confusion of significance with signification. Significance is knowledge of how an event makes a difference to your life; signification refers to the semantic content of linguistic units (DeLanda, 2011, p. 94). Enactive theorists understand mind as the creator of significance, while compositional theorists understand mind as the creator of signification. Yet both groups call their area of study cognition, so do not always clearly understand that they are talking about different activities. It is as if a group of silver miners walked into an office full of statisticians at Google, to tell them they will never discover the data they are mining without proper drilling equipment, while the data miners are incredulous that the silver miners achieve anything without programming skills. Just because two groups use the same word in their professional discussions does not imply that they even refer to the same object.

AUTOPOIESIS IN SYSTEMS THEORY, PRIMORDIAL SUBJECTIVITY

The history of the concept of autopoiesis offers anyone who researches philosophy a lesson in how to use disciplinary contexts. As philosophy became a university-based discipline, philosophical education has developed a long period of gestation. Such a long training period from student to creative practitioner is important, so that you can properly understand a concept in the conditions of its generation in an individual philosopher's writings or, like the case of autopoiesis, as the product of a larger disciplinary or multidisciplinary community. However, such a long period of apprenticeship can result in habits of thinking and expression that defer so much to established practice and a disciplinary status quo that they prevent creative progress. A concept is created in two contexts: the disciplinary framework of its developers and the problems for thought and life to which the concept is a response. Paying too much deference to the original context of a concept's creation keeps you from seeing its progressive potential when applied to different problems. Yet such a shift in application requires careful attention to the concept itself to understand what is essential to it as an idea and which of its aspects function only through the interaction of the idea and its original discipline and problems. You must be humble in the creativity of your thought without shrinking altogether from conceptual innovation. The required nuance in thought and expression is a careful balance honed only through practice.

Systems theory articulates the concept of autopoiesis more fruitfully than cognitive science, but still has significant problems of its own. It concentrates on the implications of an autopoetic body's structure and assembly, specifically its closure. In creating its thermodynamically open membrane, all the other functions of an autopoietic body become self-referring, a reflexive feedback mechanism. An organism requires an autopoietic membrane for its continued stability, but its construction also turns it away from all that is outside it, apart from thermodynamic exchange. An organism's own idiosyncratic form conditions all its activity (Maturana and Varela, 1980, pp. 50–51), which means that a system develops in isolation from all surrounding bodies (Maturana and Varela, 1980, p. 73). An autopoietic body can still affect and be affected by its bodies in its environment and surrounding bodies, so this is not an absolute isolation.

Autopoietic bodies respond to affects that occur in its own processes and its environment, but from the perspective of the responding body, these affects are mere irritations, and it always formulates its response in terms of its own peculiar internal mechanisms (Luhmann, 2000, p. 88). Niklas Luhmann calls this mutual affectivity of a body with its environment structural coupling. Coupling is essentially constituted through feedback loops, so wherever a feedback loop constitutes a body, it is autopoietic. A body's feedback mechanisms can train themselves to respond to its environment in a way that benefits it, but this response to stimulus is a minimal response to avoid catastrophic disturbance (Luhmann, 2009, pp. 147–148). The historical birth of autopoiesis as a concept in cognitive science was ultimately a disciplinary problem with including it in an ecological philosophy, but this makes a more fundamental obstacle. If feedback mechanisms constitute a body, then it is epistemically isolated from its environment, in an absolute sense. All human knowledge is inherently anthropocentric, because, as organisms, we perceive and interact with the world through feedback loops that code all input idiosyncratically (Wolfe, 2010, pp. 112–114). Understanding an ecosystem on its own terms, not our own, is rendered impossible. However, this account of autopoiesis and how feedback loops affect knowledge fails to understand the genuine power of affectivity.

The common sense conception of communication is as transmission, the exchange of some meaningful unit from one body to another. A source transmits some message from inside itself, sending that message outside itself and through its environment until its destination receives it into its own internal contents (McLuhan and McLuhan, 1988, pp. 86–87). Transmission must proceed without altering the message at all. Otherwise, it would no longer be the same message as when it was inside its source, and the communication fails if what is received is different from what is sent. However, because the unique structure of an autopoietic body conditions all its activity, the destination body must condition the message according to its own idiosyncrasy to receive it. Such conditioning alters the message, so this model of communication is impossible (Wolfe, 2009, p. 228). Therefore, systems can never communicate with genuine semantic stability, and all existent meaning is the idiosyncratic product of each body, which it generates for itself alone (Thompson, 2007, p. 57). Every autopoietic body, including humans, is alienated from each other because they cannot communicate messages with any fidelity, but only disturb and destabilize each other's self-maintaining processes from the outside. Genuinely understanding another would be extremely rare and lucky, and would never be confirmable anyway because we can receive no faithful message in return (Bankowski, 1994, pp. 258–259). From this interpretation of autopoiesis emerges a vision of the subject as an incomprehensible, isolated unit, souls like Roquentin tortured by their inevitable loneliness.

Varela himself was often frustrated by the strong isolationist conceptions Luhmann's influence brought to discourse about autopoietic systems. Mark B. N. Hansen writes that Varela once said, "Luhmann was the worst thing to have happened to [me]" (Hansen, 2009, p. 131). Yet Varela sometimes relied on the principle of closure that he repudiates in Luhmann's work. For example, Varela critiqued the idea that natural selection tends toward species developing optimal traits for survival in their ecosystem. His reason is that a system can optimize its adaptations to its environment only if system and environment can communicate in detail. Such communication, Varela says, is impossible because a species can understand the world only on its own terms. It is autopoietically closed, so there can be no genuine exchange of information between a system and its environment. A species can only muddle through, reacting to the buffeting of environmental disturbance. At best, a system can satisfice its way to a somewhat workable solution to some environmental problem, and evolution is the satisficing process of species change (Barbaras, 2002, p. 129).

The strongest interpretation of autopoietic closure implies the absolute epistemic isolation of all bodies, a vision of perception as nearly blind stumbling through an essentially alienated world that contradicts our common sense definition of perception. It is true that the only medium between physical bodies are affects, but the strong interpretation of closure presupposes that affects alone can never be detailed enough to communicate the genuine complexity of the world. Luhmann conceives of an affect as a mere disturbance or irritation. Lawrence Grossberg, for an example from a field as distant from cybernetics as literary studies, conceives affect as a simple feeling that arises when a body detects its being somehow impinged upon (Grossberg, 2010, p. 310). However, affects can achieve highly detailed communication, because the common sense view of communication is false. Communication is one body generating an affect with a specific pattern, and directing that affect to another body, which makes sense of that pattern.

No matter how the transmitted object changes, communication will remain effective as long as it maintains its organizing pattern. For example, a star releases nuclear energy as an electromagnetic wave, which I receive on Earth as neuro-electrical impulses in my optic nerves, and perceive as a green light. But I always perceive a green light when I expose my eyes to electromagnetic radiation with a frequency of 540 terahertz because its pattern remains stable across all translations of its matter. There is regularity in perception where it counts (Brigham, 2009, p. 209). Structural coupling is the process of generating complex fields of affects with continually fluctuating intensities, and bodies adjusting to these fluctuations. Affectivity accomplishes all becoming and change, all production and destruction, all syntactic pattern and semantic content in language and thought. Structural coupling interactions may be destructive collisions that cause cascading affects that break apart the bodies that created them. Affects can also constitute a self-sustaining field, a harmonious relationship among many bodies that can extend to a much larger area than the pedestrian level of the individual bodies themselves. Because the character of the field can change radically with even a slight variation in the intensity of a single affect, structural couplings can include tremendous amounts of detail (Berressem, 2009, pp. 80–81).

An organism's perceptual apparatus detects fluctuations and relative stabilities in the fields of affects that surround it, and the affects with which our apparatus directly acquaints us constitute the character of our consciousness. We can learn about affects that we do not directly perceive through empirical investigation. But whether we can directly acquaint ourselves with an affect is our problem, and can give us no absolute standard to rank or otherwise organize the various kinds of affects. Whether an affect is directly detectable depends on your perceptual machinery. Bats and dolphins can detect fields of sound affects with such detail that they can navigate their environment by this means alone. Mormyrid, or elephantnose fish, can detect electrical fields in similar detail (Morris, 2003, pp. 184–186). Humans can detect linguistic affects, visible marks, or audible patterns that articulate ideas, descriptions, and orders.

Human consciousness is itself an extremely complex field of affectivity, and the external expression of internal processes (Deleuze, 2001, pp. 26–27). Yet despite its complexity, as with any autopoietic system, your environment as a whole will always be more complex than your own consciousness. Ontologically speaking, an ecosystem exists on a larger scale than an individual organism, even with all a human's complex organismic and internal ecologies, and contains a multitude of other individuals. But there are epistemic implications as well, because an organism can only perceive what its perceptual apparatus permits. So your perception will always miss some details of its environment (Clarke, 2009, p. 53). A human cannot echolocate like a bat or perceive electric charge like a mormyrid. Beyond physical limitations, even worldly observation and action require epistemic restriction. To act, you must ignore any affects irrelevant to that action (Brown and Tucker, 2010, pp. 234–237). Such selectivity is required for simple purposive actions like grasping an object of walking through a wood. Our enormously industrial technological civilization has developed powers to analyze phenomena in terms of many aspects. But these powers of deeper analysis and finer epistemic discrimination have not removed the problem of selectivity in practical knowledge. "We perceive only... what it is in our interest to perceive, by virtue of our economic interests, ideological beliefs, and psychological demands" (Deleuze, 1989, p. 20). Expanding our storehouse of worldly knowledge and investigative techniques only makes us more aware of how much we cut away in the moment of action.

Yet these ontological and epistemic limits do not isolate a body from its environment because they constantly interact through dynamic fields of affects. While a single system will never comprehend in detail every aspect of its environment as it continually develops, it is always adapting to the dynamic affects of its environment, anticipating, and responding to change as best it can. We surf a wave of constant change comprising so many currents that our balance is always precarious. Perfection is impossible, but satisfaction and optimizing is, although we need constant vigilance and openness to the changes that inevitably come. An organism optimizes its existence by improving its perceptual and epistemic abilities. Progress is increasing the number of changes you can make in how you understand and act. Luhmann conceived of autopoietic closure as a mechanism of internal development to complexify itself as its environment continually presents more and different challenges (Wolfe, 2010, pp. 221–222). An autopoietic body, a perceiving organism, is an element of a dynamic field as well as a body with a self-defined physical boundary. It is constituted by the turning inward that separates its body from its environment, but it remains integrated with its environment as each constantly affects the other. As well as being a self, it is also a place.

My Body Is Not My World; My World Is My Body

Arne Næss conceived of self as a place in his ecological philosophy, although the concept remained underdeveloped. He introduces his concept of self-as-place to ground the ethical attitude of identification that is central to deep ecology. Identification is an alternative to contemporary consumerist attitudes of egoism that spur widespread ecological destruction. Næss never treats identification and self-as-place in a technical philosophical manner, but as an image of selfhood for everyday life. An attitude of identification is to think of the universe when you think of yourself, "connecting the individual's unfolding to that of the whole planet" (Næss, 1989, p. 163), and ultimately the entire universe. Identification systematically articulates what Næss calls the "oceanic feeling" generated in the experience of the interdependence, integration, and oneness of all existence (Næss, 2005, vol. 10, p. 94). Næss intends this conception of the self to overcome the absolute dichotomy of humanity and nature, where giving of myself to the environment diminishes me. Instead of a self that stands separate from my world, I consider the entire universe a single body, the Self, the highest possible step in a progression of self-understanding (Næss, 1989, p. 85).

The concept of identification blends theoretical scientific investigation with a lifestyle guide. Identification is a key element of Næss's personal ecosophy, his set of presuppositions for an ecologically mindful lifestyle that mixes philosophy, religion, and science (Næss, 2005, vol. 10, p. 89). He purposely denies that his ecosophy is an internally consistent philosophical system, and dismisses such systems as mere weapons of academic jousting (Næss, 2005, vol. 10, p. 105). He advocates ecosophy as a broad field of colloquially defined attitudes and tendencies of behavior, a guide for living whose principles bridge traditional disciplinary boundaries (Næss, 2005, vol. 10, p. 489). He considers such a dismissive attitude to systematicity and argumentation, a sign of wisdom and humility (Næss, 2005, vol. 10, p. 108).[4] In this way, ecosophy is Næss's challenge to university-based philosophy, an accusation of its arrogance and irrelevance to everyday life. His ecosophy's attitude of identification calls on us to understand all individual lives to be tied together in the unity of nature as a whole. Næss calls the process through which we understand the unity of all beings self-realization. Self-realization is one way worldly activity (lifestyle) and individual thought (philosophy) constitute a human's personality.

In his logic of preciseness, Næss denotes self-realization broadly speaking as T_0. He describes three modes of self-realization: ego-realization (T_1), self-realization (T_2), and Self-realization (T_3). He analyzes them in terms of their conativity, how each personality type takes stock of her own powers and imagines how her relations with other affect her powers. In this context, we can draw ethical applications from the perceptual activity of organic autopoietic systems. An organism's activity constitutes significance in its relations with surrounding bodies according to the benefits or harm those other bodies afford it (DiPaolo, 2009, pp. 14–15). Ego-realization produces an egocentric personality, which expresses a partisan us-against-them attitude that "stresses the ultimate and extensive incompatibility of the interests of different individuals" (Næss, 1989, p. 85).[5] Næss conceives that the egocentric person understands his relations with other bodies as a continuing struggle without exemption or escape. An egocentric shies away from relationships with others because they will likely harm him or benefit from his misfortune. A mature path of self-realization emphasizes the relationships that you must pursue to maintain your own life. Næss describes this focus as the self's perseverance to develop beyond an egocentric's defensive aggression. Such genuine self-realization maximizes your existing powers and develops new ones (Næss, 2005, vol. 10, p. 292). Both paths of realization focus on your own good as an individual body. The egocentric has simply blinded himself to the truth that interdependence with others often brings mutual benefit.

Self-realization (T_3) is a paradigmatically different attitude to our relations with others. Self-realization (T_2) is a nuanced, open-minded attitude toward other bodies, in which we examine relations for their benefit or harm they bring us, and act accordingly. This kind of self-realization is optimal to understand practical differences in the world. Self-realization (T_3) is an attitude of love toward all other bodies that motivates your creation of an entirely positive and loving environment. This attitude opens you to understand absolute difference. You no longer understand yourself only as an organism with needs, desires, and vulnerabilities. You are such an organism, but you are simultaneously a component of a whole that "hangs together" (Næss, 2005, vol. 10, pp. 519–520). Each part of such a whole depends on every other for its continued existence and development, and the entire ensemble constitutes a home to which each belongs (Næss, 1989, p. 164).

To illustrate this attitude, Næss describes Norway's small, isolated indigenous fishing communities, who government policy forced to resettle so they could better access state services. Although they received many benefits from resettlement, such as health care, improved infrastructure, and integration into a diverse society, the move diminished their identities because it removed them from the land with which they *identified.* The ecosystem where they lived was as much a part of who they were as any cultural heritage (Næss, 2005, vol. 10, p. 521). Their identity was more than the obviously human elements of their personalities, livelihoods, and daily habits; it was inextricably integrated with the ecosystem of which their society had been part for centuries. Self-realization (T_3) is identification with the place where you live, your conception of yourself is literally self-as-place. When you identify as your place, you make the benefit of the place literally your interests (Diehm, 2007, pp. 4–6). As your attitude of self-as-place identification matures, you expand the scope of your place, and so your self, beyond where you live: from the site of your home and its surroundings, then an entire ecosystem, a planet, a solar system, galaxy, and the universe. The most enlightened identification encompasses all (Næss, 2005, vol. 10, p. 524).

Self-as-place is an unusual philosophical concept. When you believe in self-as-place, you express an attitude that shapes your personality and moral decisions. It is an ethical concept with political power to subvert the egocentric attitudes that Næss sees as endemic to Western consumerist culture. An egotist's most difficult task is sincerely caring for others. Ego-realization is hostile and fearful toward whatever is different from himself, and what is directly harmful to the ego deserves such hostility. Even the more charitable self-realization reacts with fear and anger to what harms her. Yet the most enlightened ethical attitudes of self-as-place loves and cares for the tornado destroying her neighborhood. This attitude even loves bodies that have no obvious effect on her at all, like distant galaxies and Antarctic soil microbes, which are so different from a human self that direct sympathetic identification is extremely difficult. Næss compares such sympathy to the highest moral law in Kantian tradition of philosophy, although he takes quite a liberty in doing so. The highest moral law for the environmentalist would be a duty to care for what is different from yourself. Such a goal is the highest because it is so difficult to achieve. Yet Næss does not want his moral perspective to be a remarkable few's pinnacle of achievement, but the standard attitude of all people (Næss, 2005, vol. 10, p. 527).

Næss holds the reasonable belief that the majority of people in Western industrial societies realize themselves as egos, as an isolated, radically mobile body that has no essential connection to place in its identity. Ego-realization parallels the strongly isolationist interpretation of autopoiesis where the self is inherently and incurably alienated (Luhmann, 1990, p. 45). Luc Ferry calls such an uprooted and uprooting self the highest ethical achievement of Western liberal humanism; such alienation is for him the highest freedom (Ferry and Renaut, 1990, p. 4). However, ego-realization is not humanity's only possible subjectivizing trajectory, and certainly not the best. Many, Næss included, often look to indigenous cultures for examples of societies defined by more enlightened forms of self-realization. David Abram's work is a very clear example of the benefits and problems of this approach. His account of indigenous Australian cultures reveals a functional society of individuals operating with an integrative conception of selfhood. Yet this style of discussing indigenous cultures easily slips into essentialist caricatures of indigenous persons as Edenic stereotypes like the noble savage and the Magical Red Man (Abram, 1996, p. 93).

Only humanist presumptions make the idea that your self-conception should include physically separate beings sound paradoxical. Different ontological or religious presumptions about the nature of the universe can turn an apparent paradox into the banal truism of a pop culture phenomenon. Consider this tweet from the Dalai Lama. "Due to the fundamental interconnectedness that lies at the heart of reality, your interest is also my interest" (Dalai Lama XIV, 2011). Næss writes:

> The greater our comprehension of our togetherness with other beings, the greater the identification, and the greater care we will take. The road is also opened thereby for delight in the well-being of others and sorrow when harm befalls them. We seek what is best for ourselves, but through the extension

> of the self, our 'own' best is also that of others. The own/not-own distinction survives only in grammar, but not in feeling. (Næss, 1989, p. 175)

If we all conceive of ourselves as a place that encompasses our entire planet, sympathy for the radically different is no longer so difficult. "If your self... embraces another being, you need no moral exhortation to show care" (Næss, 2005, p. 525), because I already understand this other as myself. I am the same substance as every other body in the universe (Sowards, 2006, p. 50). Difference is trivial in the face of the ontological unity of all as whole. Because I think of myself as the entire universe, I care for all by caring for myself.

When My Self Is My World, Where Am I?

The concept of self-as-place has a lot of problems as Næss initially conceived it. To begin, there are inconsistencies between his practical motives for encouraging this lifestyle shift and his broader vision of how philosophical thinking works. When Næss was affiliated with the Vienna Circle as a young man, he contributed to their vision of philosophy as clarifying the statements of science and making them more precise (Howe, 2010, p. 370). His later use of those old precisation tools to analyze political slogans and normative statements shows that he never entirely abandoned those goals. Yet in pushing his account of self-as-place as the ground of a popular morality of sympathy and harmony, he endorses a vague principle instead of giving it rigorous content. The simple holism of his ethical concept of self-as-place blurs the distinction of subject and world, which is incompatible with a scientific conception of the human subject, which must include an account of organisms as autopoietic bodies. If we do not harmonize the scientific and the ethical, then environmental philosophy will be stuck developing moralities that are incompatible with the ontologies biological research suggests.

The first ontological problem for this ethic is uncertainty about the minimal stability required for a self. We often consider identity a stability in a world of flux, but uniting self with its ecosystem introduces countless dynamisms rocking selfhood from its intuitive stillness. Consider a corporation. It is produced and maintained by the activity of its human employees, the physical architecture of buildings it owns, and its communications infrastructure. Later, it may incorporate new buildings, vehicles, and technologies, which may interact to transform those old patterns of activity (DeLanda, 2006, p. 44).[6] A company that begins operating frozen banana stands can later build palaces for dictators, replacing its entire employee roster and changing its internal infrastructure while remaining the same entity as long as there is continuity in those physical transformations. Such continuity is the process by which the earlier activities change to the latter. Processual continuity provides the essence of a genuinely dynamic selfhood, but only at local scales. The most mature realization of self-as-place is identification with the entire universe. Yet the dynamic pattern of the universe as a whole is neither variant nor invariant. It consists of all radical creations, destructions, and transformations (huge star-factory nebulae, biospheric planets), but also all practically unchanging stabilities (supermassive black holes, enormous swaths of empty space). We can say about ultimate totality only that all that is, simply is.

A solution lies in the political convergence of ontology, ethics, and morality? The widest concept of self-as-place, the self as universe, is ontology consciously employed for political means. The goal of Næss's ecological philosophy is the creation of a society, which holds religiously the maxim, "Every living being should have the right to live and flourish" (Næss, 2005, vol. 10, p. 292). Such a society will avoid ecological catastrophe and repair damage already done. Making individuals' ontological discreteness paramount prevents my interest from ever becoming your interest unless I negate myself morally to act as another's vassal. I take the interest of another as my own when I understand myself to be integrated ontologically with my habitat and all other living beings. If all people think this way, we will settle conflicts of interest with compromise because the betterment of us all will be the paramount concern. Egotistical selfishness will have been overcome. For Næss, the daily lives of individuals articulating such a state of mind realizes Spinoza's joyful state of existence on a civilizational scale. The difference is that Spinoza described joyful existence as an individual concern, while Næss conceives of how an entire society can achieve it (Næss, 2005, pp. 528–530).

The concept of self-as-place is the key social dogma to support this noble goal. But as a social dogma, it is a fiction. It is not a fiction in the sense of falsity, because it is not a proposition, but an ontological concept about the universe upon which belief in a morality depends. Connecting the concept's ontological and moral/political aspects is very hazardous for two reasons. First, there is no necessary harmony in the interactions of autopoietic systems, but mutual affectivity, which can be symbiotic, parasitic, resonant, or destructive. This lack of ontological harmony among autopoietic systems constitutes their epistemic discord, where each system is continually catching up to the fluctuations of surrounding affects (Clarke, 2009, pp. 51–52). Genuine and intimate knowledge of another comes only from deeply focused attention. Næss declares that a society of selves that consider themselves places will relate to each other and their ecosystems harmoniously. Discord should be exceptional and strange, when it is in fact an ordinary condition of life. Second, the universe does not meet the minimal conditions for autopoietic selfhood because it is not an organismic unity, but an aggregative unity.

I can rescue the concept of self-as place for ecological philosophy, at the price of critically transforming Næss's original philosophical innovation to remove the harmonious holism that conflicts even with the model of compromise-based environmental politics he presents in his "Wolf Policies" paper. The deep ecologist's goal of identifying with the entire universe has been dismissed as oversimplified and naive (Kretz, 2009, p. 128). Such simplified holist identification ignores the reality of difference and thereby prevents us from acknowledging the moral importance of the creative proliferation of difference (Diehm, 2007, p. 9). Næss wanted his concept of self-as-place to ground a moral redefinition of your self-image so that self-interest corresponds to all-interest. This moral thinking presupposes a Spinoza-inspired ontological conception of God as the unifying principle of Nature that remains vague enough to be accepted as popular dogma. But an image of universal harmony interferes with the challenges of environmental politics when compromise is inevitable to settle conflicts of ecological interest.

Spinoza's concept of God as Nature can be an ontological organizing principle to understand the universe as a complex whole constituted through the mutual affectivity of processes, some of which are autopoietic. The concept of self-as-place works in a comprehensive ecological philosophy when we understand the oneness of an individual with its place not as a naive unity but as a complex whole. You might regard a complex whole as a unity, referring to a simple *x*, if that is all you need do at the moment. But a complex whole is a plural body, constituted from a singular assemblage of fluctuating affects whose stability is not stasis, but continuation. A body is a field of dynamic affectivity. Self-as-place understands a human subject as a place or ecosystem in which its subjectivity takes priority in analysis and practical action. I act mindfully of the interests of all bodies in the place, but I still act as a subject. "The free man acts out of virtue, that is, he performs things which can be understood from the laws of man's nature alone. This implies that the free man can be adequately conceived by himself, which in turn makes man . . . a substance. Ultimately, this implies each man is God himself" (Næss, 1975, p. 57),[7] that each of us is an expression at a singular time and place of all reality.

When you understand yourself as a place that includes all that is, you are free when you self-consciously act mindfully of the welfare of everything. You as a subject are at the centre of the universe, but only in the context of your own action. But Næss centers his philosophy of self-as-place in the moral and political domain, the absolute priority of the subject in practical action sits uncomfortably with an ecological ontology where the subject is a field of affectivity with no priority over any other fields. The solution lies in the nature of a body's interests. In mainstream moral philosophy, interests are the product of the capacity to articulate them in language, which means that only humans truly have interests (Aaltola, 2008, pp. 187–188).[8] But in Næss's thought and deep ecology more broadly, a body's actions to preserve itself, whether as a stable form or in its continuity of process, constitute the minimal ground to conclude that it has interests.

Animal organisms without linguistic abilities articulate interests as desires. Other kinds of organisms articulate them as strivings to live. Nonorganic ecosystems articulate their interests as stabilizations. Humans can study the processes by which these bodies preserve themselves, and articulate them linguistically as interests, becoming spokespersons on behalf of bodies that cannot speak (Latour, 2004, pp. 64–65).

Philosophical analysis can systematize the diverse interests of all bodies into moral norms, principles, and imperatives, and further derive specific ecologically minded political policies and laws (Næss and Mysterud, 1987, pp. 26–29). Being derived from the physical activities of bodies, these moral notions are immanent to physical existence. Næss calls this conception of interest his Spinozist ethics (Næss, 2005, vol. 10, p. 300). He generalizes over all of nature the morality of fairness in assessing and balancing interests, and upholding the interests of those who are unable to do so. Entire ecosystems, and nature generally, are held to be moral subjects (Kohák, 1984, pp. 1–6). Only the human subject can articulate this morality because only the human mind can use language, philosophy, and the science of ecology to observe, analyze, and understand all the interests of every constituent of his place.

Genuine Freedom: Beyond Active and Passive

Being himself as well as his place, the figure Næss calls "the free man" investigates the processes that constitute his place to know himself better. He observes the myriad activities, which articulate the interests of an ecosystem's bodies and relations, and make upholding those interests into interests of his own (Diehm, 2010, p. 8).[9] The ground of freedom is not any exemption to a deterministic order, as in the liberal humanist dichotomy of an active humanity and a passive nature. Activity, the flux of process, generates new singularities all the time, meaning that each change is a departure from an old order that, however minor, constitutes a free act. It does not matter if any of these fluctuations are predictable; dynamism alone constitutes freedom. Freedom is the default state of any entity.

Contrary to a humanist tradition in which humanity is the only free creature, human self-consciousness is the only way a body can enslave itself, even while freedom manifests in the dynamic flux of activity alone (Næss, 1975, p. 38). When a body is self-conscious, how you understand yourself determines whether or not you can live in the freedom proper to all dynamic bodies. Reflecting on your place in the world abstracts your thinking from your continuing existence, to examine the present flux, remember what came before, and anticipate what can be. How we conceive of what we are sets our presuppositions about what we can do. Such reflection on your ontological nature, no matter your technical skill or background knowledge, is philosophizing. If you conceive of yourself as the free dynamic body that you actually are, you are free.

But if you conceive of yourself otherwise, your presuppositions about what you are constrict what you can do because your self-conception is inadequate to your actual powers as a dynamic process. The egoistic model of subjectivity constricts you because you conceive of yourself as absolutely distinct from your environment and other people. So an egoist cannot develop powers that depend on his interdependence and connection with the world. Because an egoist cannot act intentionally on processes of which he knows nothing, he is an inescapably passive victim of their changes. Our self-consciousness enslaves us when we conceive of ourselves according to models that make us ignorant of our true freedom. Despite all of this being in Næss's philosophy, his writings sometimes ignore the conditions for a self-conscious body's freedom. While conceiving of the world as an interlocking network of dynamic movements, he often wrote that only self-conscious creatures could become genuinely active because only they could understand how one affects and is affected. All bodies that could not know themselves were essentially passive, determined by the movements of everything else in the universe. Here he follows Spinoza too faithfully (Spinoza, 1992, pp. 204–205).

Næss is concerned that the concept of self-as-place prevents a subject from closing its understanding and moral concern into itself as a ego. Egoistic control is domination, the instrumental attitude at the ethical heart of ecological crisis (Evernden, 1985, pp. 14–20). Environmental philosophy subordinates the ontological matter of what you conceive yourself as being to the moral concerns of the ecological crisis alone. This is fine if your task is the immediate practical concern of political change in the face of ecological crisis, as Næss's was. But morality alone cannot motivate genuine and complete political change. The profundity and implications of the contemporary ecological crisis make this tension inescapable. The ontological matter of how you understand yourself and your place in the universe must operate in tandem with moral concerns. Instead of an isolated unity, understand yourself as a plural entity: a multiplicity or a complex whole.

When people understand themselves as place, they will negotiate among all the world's interests whose moral standing equals their own to conserve their place's ecological integrity. The deep ecology movement works politically, morally, and ethically, but an ontological problem remains. This profound transformation in thought encompasses the scope of the universe, but happens within an individual self-consciousness, whose intentionality unifies a consciousness' phenomenal experience into a focused movement of perception and action that ignores the diverse field of its origin (Merleau-Ponty, 1962, pp. 115–116). If you let such a subsumption occur when you understand yourself as a place, environmental affects appear to you as intentions. You will continue to understand all active force on the model of human intentional action, and illegitimately personify everything in nature.

Yet this false view serves deep ecology's goal of moral, political, and social change, even as it mocks our scientific knowledge of the biosphere's profound diversity. Interests grounded in activities are mistakenly understood to be intentions grounded in self-consciousnesses, and the self is defined by the moral duty to harmonize conflicting intentions. Personifying natural bodies may work as a religious principle to organize society. Religious doctrines have involved more convoluted metaphysics, and a personified nature conforms to widely held stereotypes of the indigenous belief systems that the most embarrassing environmentalist philosophers romanticize. But it cannot work as an ecological philosophy. When we understand the complexity of the universe as a plural subject whose diverging component interests are human-style intentions, anthropocentrism, a conceptual bugbear of deep ecology, reappears.

All the interests and desires of the human population are certainly diverse and mutually inconsistent, but these are small variations compared to the wide divergence of interests among all the organisms in an ecosystem, let alone a planet. These diverging interests emerge from ecological relationships. Many relationships are symbiotic, but others are predatory or otherwise incompatible. Predator-prey or parasitical relationships are standard examples of such incompatibility, but interests conflict within the same species and organic community. Consider the New England white pine. Saplings require intense sunlight to grow, so their first generation tends to thrive in brush. But succeeding generations tend to fail because adult pines block the direct sunlight the younger ones need to develop. An older generation of the same species interferes with the interests of its own offspring (Lewontin, 2000, pp. 58–59). Conceiving of a place as a *self-conscious* complex whole requires either ignoring the conflict among an ecosystem's constituents or considering that conflict inconsequential.

Human attempts to understand nature tend to similar mistakes of personification. For example, animal rights morality understands all animals' desires not to suffer as the same kind as humans' desire not to suffer. Ontologically speaking, however, each pain affect is different from every other, even when they occur to the same person. The degree of difference among pains is greater when we contrast pain affects of different organisms, and greater still contrasting pain affects of different species. Saying that it is all the same pain ignores this difference. The pain of a human and a sheep are equally worthy of consideration when dealing with the moral and political problem of alleviating the suffering of animals. But this assessment is restricted to this problem in the domain of moral philosophy, and is irrelevant to ontological concerns.

The panpsychic interpretation of autopoiesis in the theory of the enactive mind commits the same error, equating interest with intention. Thompson takes the activity of even the simplest organisms to be self-consciously goal directed. Varela's conception of sense making as the constitution of interests through worldly activity avoids his successor's more egregious error. Sense making as the creation of significance in the world diverges from more specific conceptions of human interest. I can have an interest in preventing Arctic ice melt, with the intention of preserving humanity's coastal cities for future generations of humans. Such interest is not a worldly action, but a complex desire in thought. The activities of nonlinguistic mammals and possibly octopodae approach the anthropic model of interest as practical intention, but insects, worms, plants, fungi, and the staggering diversity of unicellular life and viruses are further from this paradigm. Nonorganic assemblages like mountains, oceans, and plastics are active in the sense of affectivity, but diverge even further from the intentional model of interest. Understanding yourself as place that subsumes ecological activity under the interests of a universal self can make you forget these ontological divergences. A subsumptive concept of self-as-place is a holist vision of the universe whose oneness articulates a subject's interests. A double naivete hobbles a subjective holist universe. It not only falls into the Edenic stereotype, but also constitutes a naive unity, an ontological impossibility.

The panpsychist interpretation of self-as-place parallels the error of believing that autopoiesis alone is sufficient for linguistic self-consciousness. Thompson is correct in identifying that autopoiesis constitutes the most rudimentary consciousness possible, but autopoietic activity alone does not constitute a self-conscious subject. Insofar as an autopoietic body is a necessary physical condition for motor and perceptual powers, one can call it a minimal self (Thompson, 2007, p. 260). Thompson simply presupposes that conativity alone is the least mechanism required to constitute significance in the world. He therefore misunderstands just how rudimentary such a minimal self is, and so believes that conativity alone constitutes human selfhood's complete range of worldly activity and self-conscious knowledge. Thompson's error was also based on his mistaken presumption that the creation of significance in the world is equivalent to the creation of linguistic meaning. These presumptions, results of his intellectual milieu in cybernetics and philosophy of mind, illegitimately bestow a human image on all forms of life, no matter their true variety.

Thompson's conceptual subsumption of all life in a human image commits the same error of anthropomorphism as Næss's moral subsumption of the universe in a human subject conceiving of himself as a place. Yet we should not take the opposite extreme to correct anthropomorphic mistakes. Subsuming your subjectivity into your ecosystem can easily lead to fatalistic attitudes. Understanding the insignificance of our subjective level relative to the ecosystemic and astronomical scale of space and history rightly reveals that prioritizing human concerns in all cases is illegitimately egotistical. Yet devoting yourself to politics requires convictions in humanity's immense moral significance, the foundation of which is self-consciousness (Næss, 1975, p. 53). The deep ecology political movement, in which Næss was an influential intellectual, aims to transform the human lifestyle according to ecocentric norms. This political context shaped his ecological philosophy through its goal of fostering activists' solidarity with the ecosystems they aim to protect. There can be no greater solidarity than the belief that person and place are one unified self. And there is incredible power in a political agitator's assimilation of an entire ecology or biosphere's vastness and diversity into the momentary scale of his own self-consciousness.

The paradoxes of thought and action reach a high intensity of tension in this context, and we should hope that they can spark some productive thought, if not resolve themselves. When times call for urgency, action must be speedy (Næss, 1989, pp. 23–24). So political expediency prevails over philosophical carefulness. The philosophical expression of an environmentalist political movement must inspire the proper humility in the ecocentric society of the future. One effective route is demonstrating the smallness of humanity relative to Earth. The concept of self-as-place condenses all the vastness and diversity of Earth in the political agitator's single self-consciousness, so she can overcome the fatalistic and ineffectual feelings that accompany a sense of smallness and take control of her situation. The paradox is that humanity becomes the measure of all to combat an egotistical society and create a humble humanity.

I would not have discussed Næss for so long in this book if I thought these paradoxes made his ideas useless or wrong. His Spinozist influence salvages his ideas from his own mistakes because such a theory can combine the strong unity of holism with the complexity of a universe constituted from relations among dynamic processes. This context lets pure difference coexist with unity, while moral action remains relevant. Næss writes:

> The nature conceived by field ecologists is not the passive, dead, value-neutral nature of mechanistic science, but akin to the *Deus sive Natura* of Spinoza. All-inclusive, creative (as *natura naturans*), infinitely divers, and alive in the broad sense of panpsychism, but also manifesting a structure, the so-called laws of nature.... The Nature with capital N is intuitively conceived as perfect in a sense that Spinoza and out-door ecologists have more or less in common: It is not narrowly moral, utilitarian or aesthetic perfection. Nature is perfect "in itself." (Næss, 1977, p. 46)

Explicitly understanding God/Nature as panpsychism commits the anthropocentric fallacy of understanding a concept describing an alien entity according to human structures; in this case, understanding God in terms of the human mind. Spinoza himself overcomes an anthropocentric account of existence because the *Ethics* describes God/Nature's infinity not in the typical sense of a negation of finitude, but as "unqualified" affirmation without restraint (Spinoza, 1992, p. 34). Yet anthropocentrism returns in Næss's own Spinozism. Insofar as only self-consciousnesses can affirm, comprehending God/Nature's infinity as affirmation implies that a human element permeates existence as a whole. Because the act of affirmation is itself virtuous, the infinite self-affirmation of God/Nature constitutes the goodness of the universe (Næss, 1975, p. 111). Spinozist thinking can also understand God as the word, voice, or face of the universe, adding to the anthropocentric imagery (Deleuze, 1990, p. 123). The idea that the universe is infinitely virtuous expresses the Edenic character of some environmental activists' images of nature, despite many real organisms' relations of discord and mutual exploitation.

Næss's Spinozist influence has an ontological flaw in addition to its ethical flaw in the anthropomorphism of virtue. The unity of the universe as God/Nature subsumes difference. Ontologically, all bodies are complex wholes, but the unity of God/Nature makes all the complexity in the universe a feature of its unity. Unified God/Nature expresses itself as processes that interact to develop difference and novelty in the multiplicity of attributes and modes. But unity remains the ontological priority, subsuming singularities into an all encompassing totality of God/Nature's infinitude. Substance expresses itself through the complexity of its modes, but unity produces complexity. Despite complexity being its product, we all too often understand difference as the necessary unfolding of a primal unity. The complexity that develops is contained entirely within its starting conditions, and so a derivative of unity. To illustrate, remember the preformationist model of biology, in vogue during the nineteenth century: it conceived an organism's entire future development as contained within its embryonic origin (Lewontin, 2000, p. 11).

Conceiving of process as unfolding a potential that its origin already contains is a teleological model of development inadequate to the actual constitution of difference. Næss understands finite existence as a plurality of forces that differ in their kinds and levels of power to affect and change other forces, including themselves (Næss, 1975, pp. 66–67). But the affirmative infinitude of God/Nature already contains all possible differentiations, no matter what their particular changes at any one moment may be. That a body has the power it does is a necessary aspect of its nature (Deleuze, 1990, p. 38). When that body is the entire universe, all the powers of every body that exists are necessary features of its nature. This teleological Spinozism treats the universe as a naive unity, that which simply is. Difference does not disappear from how one understands existence in this kind of Spinozist philosophy, because differentiation occurs on smaller scales than the universal. But the unity of necessary existence at the level of the universe subsumes difference at local scales.

The Radical Change of Process: Production

An ecological conception of self-as-place can expand that place to unify the entire universe only through simultaneously accepting the reality of production. Spinoza conceives God/Nature as the only self-producing being, and essentially so (Spinoza, 1992, p. 34). Bodies on a smaller scale are produced through conatus, self-preservation processes of maintenance and striving to perfect themselves (Næss, 1975, p. 84). "Levels of perfection are measured in relation to the strivings of each thing, there is no general measure such that man might be termed more perfect than an amoeba or a tree" (Næss, 1975, p. 96). Despite their never having any formal contact and working on what they conceive as utterly different projects, both Næss and Thompson similarly anthropomorphize conatus. Thompson conceives all conative bodies as subjects on the human model, Næss as striving for perfection with a human-like purposiveness. Thompson is explicitly anthropomorphic in his theory of conativity and mind, and is fine with all life fundamentally resembling humanity. As he is a philosopher of biology and cognitive science, he need have no concern about anthropomorphism as such. Environmentalist thinking mandates a moral injunction against anthropomorphism because it expresses the same arrogance that motivates the shortsighted development of enormous industry. Anthropomorphism of autopoietic sense making is only a problem for Thompson's philosophy because it is false.

Despite his own unintentional anthropomorphism, Næss's writing hints at a solution. He understands striving for perfection as self-realization, the aspiration of each body to be the best it can be, carried out by interacting with surrounding bodies such that they mutually empower and perfect each other (Næss, 1975, p. 98). So self-realization is not only a human movement, but the natural movement of conativity in all life. Human self-realization through how we conceive of ourselves is only the most advanced form yet of a process essential to all life, and the concept of self-as-place is its most advanced product. Self-consciousness is no perfection itself, but a tool through which we can achieve perfection. Conceiving of yourself as place perfects you. You further develop your perfection by using your self-conscious planning abilities to help perfect other bodies.

Circles of environmental moral philosophers regard Spinoza's era in the history of Western philosophy quite negatively for having originated the modernist form of the absolute dichotomy of human and nature. Gilles Deleuze succinctly describes the environmentalist beef with modernism as its ontology of the ideal alienating thought from the physical world when it conceives of self-consciousness as the source of the universe's order and meaning (Deleuze, 1988, p. 18). Deleuze analyzes how the purest form of this alienation emerges in René Descartes's mechanistic conception of the material universe. In this, he is consistent with Cartesianism's rhetorical role in environmental philosophy as Modernism's default whipping boy. Deleuze's account, however, is more nuanced historically and conceptually.

Environmental philosophy's typical Cartesian stereotype reduces Descartes's thought to its dualism of thought and matter that originates the mainstream Western view that human mind and material nature are absolutely separate. Deleuze looks deeper into Descartes's dualism to the greater problem of his reliance on mechanistic models of causation. A mechanistic universe does not produce anything, but simply moves with thorough determinism according to its structure, the laws of nature (Deleuze, 1990, p. 227). A clock, the epochal image of mechanistic causality, does not grow new springs and gears as it ages, but moves its parts according to their invariant structure. A system's structure contains its entire future development. What now is prefigures all that will be; the origin of what is prefigures all that can be. Apparent production or creativity is merely the unfolding of possibilities contained in the actual present. Nothing in the essence of a body ever changes because laws of nature transcendent to bodies entirely determine them. Yet the presumption of a clockwork universe is wrong. Processes of producing bodies is ongoing because the laws of nature alone do not determine exhaustively what a body can do.

Spinozist influence can only function properly when we drop Modernism's deterministic ontology from his conception of the unity of existence. Instead, we combine the unity that emphasizes the singularity of its constituents with an ontology of contingently operating mechanisms. Consider with Næss the Spinozist conception of the laws of nature as the structure of the necessary existent, God/Nature. The structure of nature does not determine necessarily the fields of affects that bodies generate through their mutually constitutive relations. Such relations are contingent in that the conditions of their origins do not condition exhaustively those processes' full range of possible development. Such a Spinozism that departs from Spinoza himself, who wrote, "From God's supreme power or infinite nature an infinity of things in infinite ways . . . have necessarily flowed or are always following from that same necessity, just as from the nature of a triangle it follows from eternity to eternity that its three angles are equal to two right angles" (Spinoza, 1992, p. 45). This necessity is evident when all bodies and their relations are explications of a unitary God/Nature, the primal order immanent to the universe itself. Spinoza's conception of necessity is even more strict than the clockwork image, material causation as unfolding with increasing precision the already existing details of unified being.

Ecological ontology literally turns Spinoza's conception of nature upside-down: the unity of the entire universe is an aggregate of all the smaller, local orders. The ontological primacy of locality replaces the necessity of explication or unfolding with the contingency of production. Such contingency enables the creation of singularity in its absolute sense. Genuine singularity, what disturbed Roquentin, is a singularity whose identity cannot be exhaustively described using general statements that can apply to past situations or conditions of generation. Explicative and mechanistic models of causality permit exhaustive description because any unfolding developments are entirely contained in a system's initial conditions. If you understand the precedent, you understand all that follows, because an event in an explicative or mechanistic order of production follows its precedent in a logical inference or deterministic cause, respectively. There is no existential dilemma of singularity in explicative or mechanistic causality because no genuine uniqueness is possible in this order of being.

The processes of a contingent universe produce singularities, developments which surprise surrounding bodies, and to which they must adapt and adjust. Each process is an assembly of singular affects, and the universe's unity is constituted from processes colliding and integrating. Order does not unfold from a primary unity, but is constituted historically from the relations among events as they occur. Constitutive relations can generate divergences from previously established patterns of affects, genuine novelty, because the structures of the relations that generated them do not exhaustively determine all that a newly generated relation can do. The assembly and disassembly of constitutive relations, bodies, and systems throughout the universe has a kind of agency that is utterly different from intentional human self-consciousness. All the parts of a body are mutually affective, in relation with each other and themselves, and the unity they constitute is never completely harmonious. As long as mutual affects among parts strengthen their bonds with each other, the unity holds together. But if any of the affects change in a way that weakens those bonds, it can lead to the dissolution of the body. The tension that is always present in unity facilitates flux. A body's unity is dynamic because it is always changing or open to change (Bennett, 2010, pp. 23–24).

However, the agency of assemblages is difficult to understand because it departs from so many of our commonplace presumptions about the world that follow from Modernist influence. Jane Bennett looks to the tradition of vitalism for an answer, with the goal of crafting a vitalism that applies to all matter, not just the organic. However, she cannot escape vitalist philosophy's vision of the organic as exempted from the deterministic order of the mechanical or inert (Bennett, 2010, p. 76). Instead of conceiving activity as freedom from some mechanism, an ecological ontology of the constitutive relation understands activity as the freedom of those bodies to act. The concept of freedom as the capacity to move and assemble means that freedom requires no exemption from an overall deterministic order and that all processes are free.

So the essence of the human subject is not some exceptional freedom, but reflective self-consciousness. The greatest power of reflective self-consciousness is a person's ability to change how she conceives of herself, which can transform her entire daily life. Conceiving of yourself as a place reveals how a field of perceptual affects is not strictly bound by the physical membrane of an autopoietic body, but integrates you with your surroundings, your ecosystem, and your planet. Ultimately, the integration of affective processes aggregatively constitute the unity of the universe. Our transformative power is a function of humanity's material structure, our bodies and perceptual apparatuses. How these condition our experience and thought is the topic of my next chapter. I explore a phenomenological account of self-as-place to describe how to understand and subjectivity as a pluralized field instead of a unified subject in self-conscious experience. Two key sources of this account are the works of Maurice Merleau-Ponty and Jakob von Uexküll. Transforming their phenomenologies into an ontology gives subjectivity a pivotal role in an ontology designed to support and complement ethical and political activity. We can achieve the personalized ontology that Næss sought with his concept of self-realization without the problems he encountered, and ground a metaphysical movement in ecological philosophy that links ontology and ethics.

Chapter 6

Discovering Active Nature in the Subject

The Western tradition of philosophical discourse commonly understands agency as an exclusive power of human subjectivity, the ability to form intentions and act according to them. The previous chapter describes an alternative conception of agency as the power to change, and the subject as one kind of field in a complex assemblage of interacting processes. However, the weight of a long philosophical tradition supports the anthropomorphic image of agency. It exercises a sedimentary social force of shared habits of thought. Concepts that go unquestioned for too long in a tradition calcify into presumptions, that which all right-thinking people intuitively know is obvious and true. In this case, the frozen presumption is the circular definition that only human subjects are genuine agents, because the definition of agency is derived from the activities of the human subject.

This chapter explores a phenomenology of self-as-place to build an ecological alternative to this moribund anthropomorphism, a philosophical account of both self and place as constituents of active nature. I could pursue the route of Martin Heidegger, who diagnoses contemporary humanity's ecological problem as having forgotten the ancient Greek understanding of nature as agency. Instead, according to Heidegger, philosophers commonly understand nature as inert, and humanity as the only active bodies, self-creating exceptions to a passive natural order (Kaushik, 2009, p. 80). But Heidegger only ever managed a diagnosis, and a dangerously loaded one at that. The works of Maurice Merleau-Ponty, Jakob von Uexküll, and the sciences to which they contributed offer more creative possibilities. Gilles Deleuze and Félix Guattari's assemblage theory, the subject of the final chapter, completes the task.

Conceiving of yourself as integrated with the place where you live is a philosophical inquiry with revolutionary power. Næss hobbled his articulation of this idea by letting the self subsume the place instead of exploring their dynamic tension. He made this mistake in good conscience, trying to mobilize political action directly from his philosophical thinking. In doing so, he fell into the paradox of environmental philosophy's legacy as a political movement, moving too fast from conceptual abstraction to social and political activism. As a result, he articulates a useful concept in a way that could not stand up to sustained philosophical scrutiny. This book attempts the same movement, but slower and more transparently so you might understand it better, and therefore be more successful.

This chapter explores how a self constitutes itself as a place in the ordinary activity of its daily life. My first step incorporates ideas from Merleau-Ponty's work in his later life, when he developed a conception of self as the activity of reflective consciousness. The self emerges from self-conscious thought and perception, to perceive the world while simultaneously perceiving your act of perception. Not only do we perceive the act of perception, but now we also have the scientific knowledge to understand its biological and psychological machinery, and its relationship with our environment. Von Uexküll was a seminal innovator in developing this science as animal ethology. Because perception is a worldly activity, the structure of your perceptual apparatus constitutes both your experience and your environment. All bodies that can perceive similarly structure their environment through their activity, but self-conscious perceivers can become aware of this process and their role in it. Deleuze and Guattari, along with their followers, picked up von Uexküll's studies as the empirical groundwork to develop concepts of how the activities of all organisms that share an environment shape that environment into a ecosystem of overlapping and integrated territories. Your environmental self-articulation is territory, the activity of an organism's daily life makes that organism both a self and a place.

WHAT IS PLURALIZED AGENCY?

I have so far analyzed what the conception of agency as activity that constitutes a process of change implies for the perceiving subject as an individual body. Every body capable of motion therefore has some degree of agency, the measure of which is our assessment of its potential range of activity. But a self-conscious subject's relationship with the processes around her integrates them with her own. As this happens, the agencies of the component bodies in an assemblage integrate as well. An assemblage's own agency is the product of the agency of each part, while each part retains its own agency. A subject self-consciously expresses its agency through forming intentions, and social interaction can share a single intention among many subjects as an implicit promise to act on it (Bratman, 1993, pp. 98–99). But self-conscious formulation and practical action on intentions is only one peculiar form among a vast diversity of agencies, and only an extremely general, abstract description can adequately explain without exception all the real-world assemblies of agency.

Here are the categories and variations of the different motions that assemble and integrate agency. Motion can be generated by a mechanism internal to a discrete body, as when cellular metabolism fuels an organism; or external to a body, as when a star's gravitational field pulls asteroids into an orbit, or a human carries rocks in a sack. Motion is not just physical bodies' movement, but also the fields of energy such as electromagnetism or gravitation that a body produces. Because all bodies can be set into motion, they can constitute an assemblage of moving parts through interacting with each other. Motion alone constitutes all the relations that assemble bodies in their singular identities, no matter their diversity and complexity. At the individual level of analysis, its identity is singular and its agency unique. At the assemblage's level of analysis, only individual bodies' functional role in generating the entire system matters for their identities. Singularity and functionality are relative to levels of analysis. Because all bodies are complex, we can analyze them as assemblages or naïve unities depending on what we want to learn about them. A complex assemblage can also articulate its agency over multiple levels of analysis simultaneously. An example would be an enormous ant colony, one region of which burrows through the ground of a forest, across farming fields, and elsewhere tunnels deep beneath and up through the sewers and foundations of a city (Geraud et al., 2002, pp. 6075–6079).

Understanding relations as ontologically prior to and constitutive of bodies and their properties, requires analyzing complex wholes as gestalts, bodies in which, if I can allow myself a cliché, the whole is more than the sum of its parts. Næss uses this clichéd image of the gestalt for the legitimate philosophical purpose of understanding the structure of bodies. His account is sketchy, mostly listing examples, many of which are drawn from perceptual experiences. Entering a room for the first time consists of experiences that assemble a complex whole from visual appearances of shapes and substances, novel smells, and lighting arrangements. A sonata is a complex assembly of notes articulated over time, and a painting of lines and colored patches existing concurrently on a canvas (Næss, 1989, pp. 58–60).

Reciprocal relationships among bodies where no one element takes prominence over any other constitute a gestalt. The concept is a useful guard against holist subsumption. It also helps us understand the integration and interdependence of a body's micro and macro scales, and processes that operate across multiple scales simultaneously. You cannot adequately understand a body without investigating its parts, nor should you investigate only its parts and their relations because some activities are generated only at the level of the whole. "Gestalts bind the I and the not-I together in a whole" (Næss, 1989, p. 60). Processes at the forefront of my consciousness in everyday experience are united in their activity with bodily and environmental processes that act upon and integrate with me. The self-as-place is a complex field's gestalt agency. Understood this way, I need never be aware of a benefit to me for it to exist. "Joy becomes, not my joy, but something joyful of which the I and something else are interdependent, non-isolatable fragments" (Næss, 1989, pp. 60–61). This is Spinoza's concept of joy as a body's state of being empowered by an interaction.

Gestalt assemblages are essentially processual. Næss often formulates his explanations of reality's processual nature as frustrating paradoxes, despite a strong tradition of process ontology in Western philosophy through the works of Alfred North Whitehead, Henri Bergson, and Gilles Deleuze. Paradoxes normally arrest thought by presenting contradictories that cannot be reconciled on their own terms. Zeno's paradoxes are the best-known example. Paradoxes can be productive, however, when they provoke us to change the logic of our thinking so that what were once irreconcilable contradictories become contraries, or even ordinary, barely controversial sentences. Yet this is typical of how Næss presents paradoxical ideas of process that animate the concept of self-as-place: "The individual selves are processes or aspects of processes, always changing, but always showing an important, limited continuity and permanence" (Næss, 2005, vol. 8, pp. 335–336). He simply presents the self as a juxtaposition of flux and permanence. The sciences of phenomenology and ethology offer us the conceptual tools to understand the permanence in flux of the self in a process philosophy.

Phenomenology: How We Know What We Must Become

Building a phenomenological perspective conscious of itself as a place requires understanding the differences and similarities between conscious and nonconscious agencies. Agency is no exception from a mechanistic material order, but an aspect of materiality itself. One central motivation of the early twentieth century's phenomenological philosophy was to combat the conception of matter as inert, valueless, mere stuff (Elden, 2005, p. 817). The inadequacy of understanding matter as passive stuff to be manipulated at will inspires the ecologically focused revival of vital materialist philosophy today (Bennett, 2010, pp. 31–32). Like these concepts of the vibrant breath within matter, Næss opposed the absolute dichotomy of inert nature and active humanity. Yet Merleau-Ponty's gestalt phenomenology overcame the metaphysics of the absolute dichotomy decades earlier than

Næss and Bennett. Merleau-Ponty used the concept of gestalt in his early work, *The Structure of Behaviour*, as a bulwark against reductive materialist accounts of reality that understood all matter, including human thinking and perception, as inert mechanism (Verley, 2006, pp. 36–37). He opposed such a conception of nature until the end of his life. His *Nature* lectures of the late 1950s reject the dualism of unified, deterministic nature and a humanity whose exemption from nature permits us to dominate it. He instead develops a more nuanced concept of active nature (Merleau-Ponty, 2003, p. 113).

The other influential phenomenological philosophies of Edmund Husserl, Martin Heidegger, Jean-Paul Sartre, and their followers conceive of freedom as a body's being able to determine itself. No matter the intricacy of its mechanical complexity, an inert mechanism does not possess freedom in this sense. This dichotomy of self-determining and inert is the central duality of Heidegger's *Dasein* and *Vorhanden*, or Sartre's for-itself and in-itself. Husserl described the dichotomy as the difference between that which can and cannot be determined (Husserl, 1970, pp. 46–48). Such freedom grounds the ability to think philosophically at all. However, these dichotomies prevent an adequate phenomenological conception of gestalt. There is in perception a difference of kind between the perceived and the perceiver, because the perceiver organizes the material of perception. Only a human (or *Dasein*, or being-for-itself) unifies the accidental aggregations of inert existents into a single perceptual field of qualitative forms.

Phenomenology gives perception priority in philosophical analysis. Each potentially sensible quantum in the world is discrete in perception, and so not essentially related to any other body it encounters in the present or the past. But perception unifies these discrete elements into a coherent field (Carman, 2009, pp. 631–632). Material existence, when conceived as the inert matter on which perception works, is an aggregation of quantities. Perception and whatever action follows from it constitute the qualitative features and unity of existence. Only in a phenomenological subject's organizing activity unifies the happenstance processes of an environment into a single perceptual field (Sehdev, 2009, p. 235). You can understand this idea as perception making an internal picture of an external environment. So the only order in the world would be its mental reproduction. But such representationalism has never been acceptable to phenomenology.

Phenomenology instead conceives of perception as a subject reaching into the world around him and ordering it according to his thought and his perceptual apparatus. The discrete objects that compose the world find their essential relations through a subject's perception unifying that discreteness into a whole. The concept of body becomes essentially superfluous to this philosophy of perception; all that matters are the disordered objects and the ordering subject. The dichotomy of humanity and nature is merely transformed into a dichotomy of order and disorder. The framework of the absolute dichotomy remains. The subject occupies the superior pole in the order of being because it can control itself and the world. Objects, take the inferior place because they are entirely passive before the control of a perceiving subject. Ultimately, this concept of subjectivity alienates a subject from everything other than itself, because the subject's ability to perceive exempts it from a universe otherwise wholly determined and inert (Mann, 2005, p. 56). Being unfree, nature is ethically and morally irrelevant (Merleau-Ponty, 2003, pp. 9–10). You can still know the world, but you are alien to it, your own subjectivity being the only source of the world's significance, the embodiment of ethical solipsism (Evernden, 1995, pp. 127–128).

Merleau-Ponty's thinking takes phenomenology beyond its articulation of the absolute dichotomy. Human agency is not exempt from nonconscious reality's inevitable passivity, but begins with material motion itself. Each motion is a singular flux colliding, converging, and diverging with many other movements, whose activity constitutes its identity. A motion "is not even a trajectory that will be, but a trajectory that is going to follow. It is the grasp of the immanence of what is going to follow in what has already begun" (Merleau-Ponty, 2003, p. 154). Knowledge of the present is also knowledge of the future, because the present state of affairs is a condition of future development. A motion's agency is its immanent power and order through which it articulates the trajectory that constitutes its singularity.

Therefore, nothing other than the motion itself determines what it will eventually become.

A process is a complex assemblage of interrelated and mutually affective motions. A human organism is a process that develops according to its own powers and structures no more or less freely than any other process. That structure is both temporal and spatial, so an organism is also its history as well as its physically bounded body. This concept of freedom overcomes the ontological absolute dichotomy whose every form entails the alienation of humanity from nonhumans: mind/world, for-itself/in-itself, or *Dasein/Vorhanden*. Each dualistic category of being is alienated from its opposite because membership in one category excludes any possibility of convergence or overlap with the other (Merleau-Ponty, 1968, p. 47). Mind is not an ontological category equal in status to a world against which it stands in opposition. It is a peculiar power of human bodies. A human is the only kind of body that, to the best of our current knowledge, can think abstractly about itself and its surroundings. Subjectivity is a process just as integrated with the world as with the body whose perception generates it (Merleau-Ponty, 1962, pp. 474–475).

The only dualism that eventually survives Merleau-Ponty's criticism is the title of his last work, *The Visible and the Invisible.* This is not another iteration of the absolute dichotomy, but a matter of epistemic accessibility: what can be perceived, and what must exist for there to be perceivers and perceived. The perceivable, perceiver, and the imperceptible conditions of perception are integrated in a single gestalt field, a unity constituted from dynamically intersecting and mutually affecting differences (Merleau-Ponty, 1968, p. 195). Merleau-Ponty aptly calls the volatile flux of complex processes "wild being" (Burkitt, 2009, p. 327). This turn of phrase alone evokes so much that is essential to environmentalist morality. Being is wild, properly dynamic, and in continual flux; humanity reduces these patterns to resources to be exploited with its brutal technology (Coleman, 2010, pp. 84–86). And once again, the simple dualisms of villainous humanity and pure, innocent nature slip into our thinking like a cigarette between absentminded fingers. Humanity can never justify its continued existence if we do not give up this tumorous idea.

The concept of wild being is better understood ontologically than morally, where a plurality of processes collide to constitute all the bodies of the universe. Among these processes is a human subject. Merleau-Ponty drifted from the phenomenological tradition in his last works, but he was still using its tools so that we can describe ourselves as such processes. Your experience of your continuing generation is a product of your generative processes and their conditions. So you cannot experience directly the processes of your own generation. But direct experience is how you investigate the processes and conditions of your body's generation. Merleau-Ponty critiques phenomenology in its Husserlian version, according to which we can refine our experience through the eidetic reduction to reveal the essence of being itself (Merleau-Ponty, 1968, pp. 178–179).[1] The priority of experience is epistemic, and phenomenology is at least ideally a method of overcoming prejudices and presuppositions that become entrenched in how we understand reality.

Merleau-Ponty's descriptions of simple experiences are starting points and guides for a philosophical account of how a plurality of processes constitute a self-conscious body in its apparent unity. A human body is a "sensible sentient" organism, able to perceive itself as well as its surroundings in remarkable detail (Merleau-Ponty, 1968, p. 137).[2] To illustrate sensible sentience, Merleau-Ponty frequently returns to the experience of a person's two hands touching each other. "The human body is, for the human, not the stand-in for or lining of his 'reflection,' but rather reflection in figural form (the body touching itself, seeing itself), nor is the world an inaccessible in-itself, but 'the other side' of his body" (Merleau-Ponty, 2003, p. 268). Perception is a process that physically connects processes that occur inside an organism with processes that occur outside. The inside and outside of a body reciprocally touch each other. The inside touching the outside is also and equally the outside touching the inside.

An epistemic implication of sensible sentience is that knowing and known are inseparable insofar as both participate in the process of perception. Processes outside a human body produce and alter inner processes, and inner processes produce and alter outside processes. The interaction of internal and external through physical affectivity is how an autopoietic system interacts with its environment. Merleau-Ponty calls this continuing interplay of processes moving inside and outside a perceiving body "chiasm" to indicate its symmetry, where no one side of the process gains absolute superiority over the other. Affectivity is mutual, so perception generates a unified field, a complex whole in which the processes of perceiving and the processes that are perceived are constitutive parts. Touch is the sense that most clearly illustrates the parity of subject and world in experience. While vision is the sense most frequently used in illustrations of perception throughout the history of Western philosophy, we experience our visual fields at a remove. Vision displays the bodies of the world at a distance; touch is inevitably direct contact (Merleau-Ponty, 1968, pp. 131–133). Self and world become a single complex body.

A human body's field of perceptual affectivity constitutes this entire lived space. Merleau-Ponty's term is "flesh," to indicate that the entire field, not simply the organic body itself, is in some fashion alive and active (Merleau-Ponty, 1968, p. 127). The word also suggests the physicality of organic life, synthesized not only from acts of perception, but from flows of water and viscera structured into relatively stable patterns by complex articulations of metabolic activity (Deleuze, 1994, p. 73). A subject is not only a field of perceptual activity, but as the physical assemblage of the organism itself. So all bodies with which you interact constitute you as visceral material: the ground on which you walk and the air you breathe, trees and buildings, which you can explore, whether or not you bother (Kretz, 2009, pp. 124–125). Understanding the integration of subject and world as flesh, an active field of affects, is irreconcilable with phenomenology's general goal, at least as it remains loyal to Husserl, of revealing the necessary universal structure of any possible subjecthood, such as a transcendental ego. Concepts of *the* Other are similarly rejected, because there is no necessary structure to the encounter of two or more subjectivities. There are processes of perceptual activity, and bodies that are perceived, or at least perceivable, because they are integrated into that perceptual field. The concept of the flesh enables philosophy to cast aside any concepts of the subject or the object's essential nature. We need not even demarcate subject and object because some assemblages can blur together perceivers and what they perceive. Your activity is always worldly because you are always a constitutive element of your world (Madison, 1981, p. 102). Rather than a dichotomy of the Self and the Other, we each exist as fields of selves and others. As Merleau-Ponty writes, "Les uns pour les autres et non pas seulement l'un pour l'autre," or "Many ones for many others, and not only the One for the Other" (Merleau-Ponty, 1968, p. 81).[3]

The Subject Emerges from the World

Merleau-Ponty retains a focus on the subject from his phenomenological heritage, which helps understand how an ecological ontology can comprehend human subjectivity's unique power of reflective self-consciousness. He tried to overcome this subject centricity in his last years, developing concepts of existence as a field of affectivity, and the subject as one set of processes among many. But he died before completing the transition, if it was even his goal. The distinction of visible and invisible is between the daily life and maintenance of a subject, and the generation of a physical subjectivity. It is a useful dualism of worldly processes that constitute the subject, and the conditions of their existence. These are the public and the private aspects of the self's unity as a place. The visible is what a subject perceives, and the invisible are the processes that constitute the body, the conditions for there being any perceptual activity at all.

A body's perceptual activity constitutes a field of motions that interact with other fields of motion around it and within it, what Merleau-Ponty calls the entire lived space of an organism (Merleau-Ponty, 1968, p. 135). A field of affects is never static because it requires continual activity to maintain its structural stability. This activity confirms our lessons from autopoiesis: maintaining a stable structure requires continuing activity. Nowhere in these concepts is philosophy's traditional dualism of subject and object, which is only an obstacle to understanding self as a field (Merleau-Ponty, 1968, p. 137). The dualism of the visible and invisible, or perceiving and what constitutes the perceiver, is better suited to an ecological philosophy than a dualism of subject and object. The relationship of our perceptual acts and the conditions for a perceiver's existence provides a framework to understand subjective experience, a subject's integration with the world in which we live, and the processes that constitute that integrated field. In Merleau-Ponty's words:

> The body unites us directly with the things through its own ontogenesis, by welding to one another the two outlines of which it is made, its two laps: the sensible mass it is and the mass of the sensible wherein it is born by segregation and upon which, as seer, it remains open. It is the body and it alone, because it is a two-dimensional being, that can bring us to the things themselves. (Merleau-Ponty, 1968, p. 136)

His phenomenological heritage leaves him concerned about how a subject can be open to the world and other subjects. A major sticking point in Husserlian phenomenology is the problem of the ego's relation with the Other. Yet Merleau-Ponty gives us the tools to escape any general epistemic skepticism about our relation with other selves or the wider world. Perception is a process that originates within us, that constitutes a field of perceptual affects, which integrates the perceiving body with its surroundings (Merleau-Ponty, 1968, pp. 38–39).

We can be flexible when we determine what processes fit in each category of the visible and the invisible. We rarely perceive our metabolism or neurological processes in our daily lives, though we can observe them by technological means. But the conditions of our perceptions are not only invisible in this workaday sense. The visible includes the pedestrian situation of my body in my current surroundings, with a particular focus for phenomenological analysis on whatever surroundings, thoughts, or memories I am paying attention at the moment. Immediacy's companion, the invisible, are all the processes whose activities have constituted my current situation across every level from the scale of the quarks to the cosmos. The material condition of every visibilities existence is its vast invisible history.[4] Perceptual activity is the only means to constitute a perceiving body, and perceptual activity cannot exist without a perceiving body (Merleau-Ponty, 1968, p. 9).

At first glance, this may seem to be a paradox of self-production: a body cannot produce itself, because it would have to have already been produced in order to do so. However, the concept of autopoiesis shows that self-production is not a genuinely destructive paradox in the case of a perceiving body. A condition for there being a perceptual field is the separation of a perceiving body from what Merleau-Ponty calls "the mass of the sensible," and what Evan Thompson called the "chemical soup" from which an autopoietic body emerges through constituting a boundary (Thompson, 2007, pp. 98–99). The body that such a process produces never stands separately from it, but is instead a continual and inseparable element of its activity. The arrangement of the parts themselves initiate transformative and generative processes that constitute the physical unity of the parts in a stable relationship. Dropping the English term for such a process and adopting the French better describes what happens: auto production (Guattari, 2008, pp. 91–92). The affects that constitute a perceiving body's boundary are the same affects that constitute its ability to move and its rudimentary perception. Subjectivity, a field of perceptual affects, distinguishes itself from a background that is featureless in comparison, its own constitution producing a novel difference (Deleuze, 1994, p. 28).

Do not presume that subjectivity broadly speaking is inherently similar to human subjectivity. Human subjectivity's peculiar nature lies in its ethical and moral dimensions, and the unique nature of the human brain. The size and structure of the human brain matter less than its process of development, the becoming of the human brain, particularly our neonatal neural indeterminacy. For each of us, many of the synaptic connections of our brains have not yet formed at birth. Our synaptic nets achieve their densest concentration and complexity through our physical interaction with our environments. Worldly action, generating and complexifying a field of perceptual affects, literally builds the human brain (Allen, 2004, p. 65). This integrates a human subject more intimately with its environments than any other organism. We generate a field of perceptual affects that profoundly transforms the world. With our mere movement through the world, land becomes territory and objects become artifacts.

Any species of organism's characteristic field of perceptual affects generates its own environment by creating new relationships among bodies, and so new significances and identities for everything involved. So humanity's environment overlaps and integrates with that of sparrows, pines, goats, spiders, *Echerechia coli*, and so on. But the peculiar plasticity of the human brain results in a new order of worldly meaning than has existed before. As a human develops its plastic brain, it does not merely explore its world. We create technical artifacts to transform the world and our brains simultaneously. We have sharpened stones and fires for most of human history. Only the mere moment of recent centuries have built our massively complex, physically devastating enormous industry. Several non-hominid species independently developed the multifaceted use of tools such as apes, monkeys, crows, and parrots (Morris, 2003, pp. 260–265). But none of these other animals have the neural plasticity that so deeply integrates technology with thought.

Brains as plastic as humanity's constitute a technological ecology that renders humanity categorically different from all other kinds of organism (Skakoon, 2008, pp. 46–47). The human environment is a specifically technologized ecology, and no other organism transforms its environment through technology as a matter of its neural development. We make the world and our brains technological as an ordinary part of our physiological development. Because humanity is so widespread over Earth, there is little on Earth that has not been enfolded into the human environment. Even the seabeds underneath the Antarctic ice, and the farthest reaches of that continent's mountains and volcanoes are the subject matter of routine human scientific activity. Barely one hundred years after Shackleton, there is an aerobics studio in Antarctica (Herzog, 2007).

You might think by now that I have contradicted myself. Having spent the entire book up to this point arguing that an absolute dichotomy of human and nature is false, I am now arguing that humanity is a uniquely technological creature of a wholly different order of being than everything else on Earth. The romanticized Edenic image of wilderness common in environmentalist moral philosophy may have a point, after all. If the human environment is uniquely technological, then only ecosystems free from any human activity throughout history can have escaped technological transformation.

There are no more such ecosystems (Rolston, 1994, p. 72). But this fact does not alone imply an absolute ontological dichotomy of humanity and nature. Having identified the unique plasticity of the human brain, we must admit that technology truly is the essence of humanity. But as one pole in a dichotomy with nature, all it opposes is that which is not human. There is boundless diversity in this category. There is no essence of the nonhuman. A blade of grass, a wombat, a paramecium, a gas giant planet, and a mushroom share no common essence. I have not identified some eternal ontological category, simply the peculiar quality of one unique creature among many.

Humanity's unique divergence from everything else is our emergence as the technological organism. Yet this does not exempt us from the natural order because technology is itself a natural development. The plasticity of the hominid brain is an evolved trait, which constitutes the conditions for our peculiar kind of perceptual field. The echolocative organs of bats or the electro-sensitive organs of the elephantnose fish are natural traits that constitute their peculiar kinds of perceptual affectivity (Morris, 2003, p. 184). Hominid neural plasticity is one more peculiarity of perception in a world of diverse peculiarities.

From the beginning of the *Homo* genus, the brains of its various species became increasingly plastic, developing their complexity through worldly interaction long after birth.[5] At least since the period of *Homo erectus*, hominid life has involved the use of tools and inventions. Artifactual activity shapes the territory in which early humans lived, and the physical comportment of their bodies, the very shape of how their subjectivity articulates itself. Human tools are minutely manufactured, as shown in pre-*sapiens* hominids' careful craftsmanship in constructing their stone tools. We can make any facet of our environment into a tool through mindful, intentional focus and exploration of what surrounding bodies can do. As we become habituated to a tool, we handle it as if it were part of our own bodies. Every tool we make expresses how our integration with the world shapes the very physical infrastructure of thought. This is true for a simple tool like a stone blade, or a highly complex tool that requires enormous infrastructure to build, maintain, and use, like a car (Dant, 2004, pp. 73–74).

Evolutionary and archaeological sciences were not the only knowledge disciplines to discover human subjectivity's profound integration with its tools. Merleau-Ponty's phenomenological and psychological investigations independently discovered this relationship, although he did communicate with archaeologists specializing in prehistoric times. His investigation reveals the plasticity of the artifact itself. An artifact's identity is constituted through its physical construction and its interaction with humans once it is built, which leaves it open to some degree of flux through human use (Merleau-Ponty, 1968, p. 75). An artifact can be used in multiple ways, limited only by a human imagination's flexibility in the application of its tools (Merleau-Ponty, 1962, pp. 7–13). Humanity's neural plasticity integrates our subjectivities with nonhuman bodies through the mutual affectivity of a developing human brain and the tools an organism crafts and uses with careful attention. Worldly technological activity constitutes fields of artifactual affects that integrate humanity with its environment in mutual affectivity.

Nowhere in this account of how humans naturally generate a field of artifactual affects in technological development and use is there a moral judgment. Moral judgments on technology are appropriate with considering topics like the destruction of other species and industry's catastrophic transformation of many ecosystems. Enormous industry has brought technological destruction and catastrophe to his highest intensity, major examples being climate change and the destruction of habitats and species. A major debate in environmentalist moral theories is establishing precisely what principles, duties, and obligations arise from our industrial powers (Palmer, 2009, pp. 587–603). Yet humans have been causing the extinction of species since their spread from Africa around the world; only now has our ecological destruction risen to the forefront of our moral deliberations.

Mainstream Western philosophy associates the technological with a departure from nature. Humanity having developed its destructive ways as a consequence of that departure, you would think that the morally good path would be to renounce technology and return to nature (Michael, 2005, pp. 50–51).[6] But the implications of humanity's neural plasticity show that technology is no radical departure from nature, but a natural evolutionary development of the hominid brain. We never left Edenic nature because there never was any Eden to leave, and our artifactual powers are entirely natural anyway. The modern ecological crisis is constituted from a long, complex series of increasingly catastrophic practical mistakes in our natural technological development. Environmentalism implies no misanthropy, but is a call to admit and repair our civilization's biggest mistake.

Strip away the idea's mythical features and the return to nature is a societal shift to activities that enforce ecosystemic stability. Much environmentalist discourse takes for granted that stability is the healthiest state of an ecosystem. But asking questions about an ecosystem's health, history, and stability makes the mistake of discussing ecosystems as if they were organisms. An ecosystem's boundary is not a membrane, but a region of space in which a qualitative shift in types of organisms occurs: an ecotone boundary where jungle becomes grassland, savannah becomes desert, or desert becomes farmland (Browne, 2007, p. 18). Like complex organisms, an ecosystem's constituents are smaller organisms. But the systematic relations among the parts of complex organisms transform their simpler constituents into organs, managing their activity in purposive relationships. This does not happen in ecosystems, where the whole does not overwrite and control the powers of its parts. Rather than the whole ecosystem guiding its constituents into purposive functions to maintain stability, the divergent activities of its constituents continually transform the ecosystem. Whatever stability an ecosystem has, lies in its periodic repetitions of fluctuations whose dynamic intensities do not approach singular catastrophic thresholds (DeLanda, 2011, p. 68).

Yet a mainstream line in environmental moral philosophy holds that each ecosystem has its natural stability to which we must return it. Such a return may take a long time, perhaps centuries or millennia, but its return is our moral imperative (Rolston, 1994, p. 76). However, the end of a destructive technological process will not automatically produce a harmonizing, stabilizing natural process. There is nothing about the origin of a process in human industry that makes it categorically different from one whose origin is in an ecosystem in which humans have never been involved. An industrial process, considered in itself, may cease, but its subsequent effects and consequences may continue long after stopping. Earth has no system restore switch to correct the aberrant processes of human industry after they stop, because they are processes just like any other, and articulate themselves through ecosystems like any other processes of planetary scale (Ward, 2009, pp. 130–132).

The typical environmentalist moral judgment is that a technologized ecosystem is corrupted and no longer natural, that technology converts flourishing, diverse life into mere resources, consumable commodities (Rolston, 1999, pp. 146–147). But humans are not the only consumers, the only creatures who treat particular other bodies as resources. Eating makes an organism into a resource. Chains of mutual consumption are a major constituent process of ecosystems. Predator and prey populations fluctuate depending on how much has been eaten and how many still want to eat. Any predator makes resources of its prey; usually with less bureaucracy than contemporary enormous industry, but the basic movement is the same. The ecosystemic fluctuation of populations sometimes tends toward stable periodic flux, but irreparable catastrophe can occur when consumption rates pass a critical threshold, and after such an event, it is extremely unlikely that the ecosystem would return to its previous state (Schaffer and Kot, 1985, p. 348). "We do just what other species do, only much more efficiently," hunting species to extinction as we make resources of them for food or industry (Michael, 2005, p. 51). I can recognize that efficiency with ironic praise. Human technology has developed ways to destroy things more rapidly, quickly, and theatrically than few other natural phenomena. Consider the efficiency with which the Aral Sea, once among the largest freshwater bodies on Earth, was transformed into a near desert over mere decades (Precoda, 1991, pp. 109–114).

Humanity's most remarkable power is the technologization of our environment. There was never a departure from nature, only a ramped increase in the intensity of consumption as we developed technological organizations that increased short-term efficiency through the uniformity and routinization of production (DeLanda, 2000, pp. 83–86).[7] Technology itself should be neither morally damned or exalted, its existence itself neither good nor evil.[8] It is a mode of existence that allows humans to thrive within the means available in our territories, or to overtax our resources to the point of catastrophe, just like many other species have done and will continue to do. Facing survival in a harsh world, there is only the question of whether some particular field of artifactual affects is sustainable or ultimately self-destructive. A way of life is sustainable if it can continue without rendering its territory unliveable for its practitioners. Enormous industry eventually makes territory unliveable for humanity.

The question remains as to what is the best way to understand how these fields articulate themselves in the world. The human environment, no matter how much our technology may dominate its surroundings, is not separate from the environments of other bodies. If it were, then one would not perceive the world, but a series of hallucinatory imaginings (Burkitt, 2003, p. 320). Each body constitutes a different field of subjectivity affects, creating a different environment for each field. But these fields all exist in the same physical region, interacting over a shared plane of space and time. Their interaction physically constitutes a gestalt ontology of subjectivity.

A Subject's World: Ethology and Territory

The ideas of Jakob von Uexküll provide a good starting point to understand how one perceiving body constitutes its environment, while all these perceiver-constituted environments interact with each other. The core of this question is the old philosophical problem of whether one can be genuinely certain that other minds, or in this case other perceivers, exist. The framework and implications of this problem cut across all four domains of philosophy. Perception constitutes a field of subjectivity, the activity of an organism in relation to its environment, which Uexküll calls its *Umwelt*. Uexküll himself describes the *Umwelt* as a total closure of the perceptual field, like a bubble around each organism. But we should not take this image too literally, as it would invalidate many actual implications of Uexküll's idea. In a field of subjectivity, a perceiving body and its environment are neither isolated from each other nor inseparable in a simple unity, but two mutually affecting poles in a complex structure of relations and feedback loops (Buchanan, 2008, p. 22).[9] A field of subjectivity is a process: a body constituting itself and its environment through mutual activity (Merleau-Ponty, 2003, p. 209). Uexküll's work and the ethological research that followed him explores the constitution of the perceptual fields that are part of this process, how an organism articulates itself in the world, consciously making itself (Deleuze and Guattari, 1987, p. 41).

The relation of a perceiver and a perceived constitutes the field of affects of the *Umwelt*, but this field is not closed because an *Umwelt* itself can be perceived. We can study an organism's movement and perceptual apparatus to discover what kinds of worldly affects it can sense and how those sensations inform its movement. Sensing is not the same as being affected. I cannot sense ultraviolet light, but it can affect my body, burning my skin, and mutating its cells into cancerous melanoma. An organism can sense an affect when it has a sensory organ like an eye, an ear, or a collection of sensitive neurons that can translate an affect into a neuroelectrical pattern. For example, an eye translates an affect of electromagnetic radiation into an affect of neurochemical activity that has a parallel pattern. To sense is to respond in this translative manner to affection.

The translative nature of perception is the ground of communication, the preservation of practically important patterns of affects across multiple media in environmental and perceptual fields. The media of affection may change, but successful communication occurs when the pattern of affection remains constant. Studying every environmental field of affects that an organism can detect lets us understand its *Umwelt* in worldly action, the complete range of possible movements in the structure of its phenomenological experience. Describing an *Umwelt* displays the ontological conditions for an organism to articulate its epistemic capacities. We do not understand another organism's phenomenological perspective through imaginative mimicry, as in Thomas Nagel's famous demonstration that it is impossible for a human to know what it is like to be a bat. Experiencing another *Umwelt* is not necessary because we can measure and understand all the affects that a creature can detect and respond to by studying its physical perceptual apparatus.

Uexküll's central example is the female tick. Its perceptual apparatus can detect only three types of stimulus, each to varying degrees of intensity. Only three kinds of perceptual affectivity constitute a female tick's entire field of subjectivity. After mating, she climbs to the highest, least shaded branches of a tree, guided by her skin's photoreceptivity, where she waits motionless, possibly as long as 18 years, for the specific scent of butyric acid in mammals' sweat. When the scent of butyric acid reaches a particular intensity, the tick drops into the hair of the passing mammal, and then seeks the hottest nearby place to reach its skin. She then bores into the mammal's skin to consume its blood. A tick does not taste blood; all that matters is that the blood be of just the right temperature (Buchanan, 2008, p. 24). Its entire phenomenal field, or *Umwelt*, is "defined by its gravitational energy of falling, its olfactory characteristic of perceiving sweat, and its active characteristic of latching on" (Deleuze and Guattari, 1987, p. 51). Empirical study has shown that these are the only affects it can detect perceptually.

With such a simple perceptual apparatus and so few movements in the structure of its possibilities, Uexküll wonders whether we would better understand the tick as a machine rather than an animal. But there is no difference of such a kind between the simple subjectivity of a tick and the highly complex and nuanced subjectivity of a human (Uexküll, 1956, pp. 24–25). The only difference between a tick and a human is in the content that each species of organism's fields of perceptual activity can generate as they integrate with their environments. A subject's experience is only as complex as its possible perceptions, and an organism's perceptual knowledge depends on the physical structure of its perceptual apparatus. In the tick's case, that knowledge is the proximity of butyric acid, the height it climbs and falls, and the heat of the blood it ingests. Those are all the affects of which it can be aware, because those are all the affects that it has the power to detect in its surroundings. Its activity may, however, generate many affects of which it can never become aware, such as the pain of a bitten animal, or the spread of lyme disease.

The activities of all organisms constitute paths for other organisms that share its territory to follow, a tendency to interdependence that produces an ecosystem. Organisms' perceptual activities constitute these paths, sometimes intentionally but more often not. One example of such interdependence is a bee and a flower. A bee must find nectar for its hive, and its availability in a flower affects a bee's movement possibilities and tendencies. Bees tend to interact with flowers because flowers are the source of nectar. Meanwhile, a flower produces nectar through its interaction with bees, which spread pollen across many flowers, facilitating the plants' reproduction. Each interaction with a pollen-covered bee constitutes reproductive possibilities for a flowering plant (Buchanan, 2008, p. 33). This is a case of mutual benefit in symbiosis: interaction among organisms that not only constitutes the possibilities for their motion, but also materially benefits those organisms directly involved. When a relationship produces a mutually beneficial practical difference, that relationship tends to recur. The repetition of beneficial symbiotic activity tends to optimize that benefit.

All interactions among organisms constitute their possibility structures, whether or not there is mutual benefit in symbiosis. Uexküll illustrates this relation between action and interaction with the example of a spider and a fly. A spider successfully preys on a fly by building a web that must catastrophically constrict its possible movements: once trapped it cannot move, and once eaten it no longer exists. Achieving this deadly constriction requires building its web to facilitate a fly's movement into it. A spiderweb must constitute more possibilities for a fly to move in the direction of the constriction than away from it (Uexküll, 1982, p. 66). Successful predation manipulates interdependence relations among organisms to craft the possibility structures of their worldly activities (Buchanan, 2008, p. 34). The technical biological definition of symbiosis is any relationship of organisms living together, and these relationships need not be beneficial. Symbiosis can include mutual benefit, negligible benefit or harm, parasitism of one on the other, or mutual parasitism (Morris, 2003, p. 335). Relations constitute not only processes, but possibilities for those processes that, without those relations, would never have existed.

Territorialization processes build the physical structures for the movement of bodies in space (Deleuze and Guattari, 1987, p. 315). The agency of organisms continually generating their overlapping and colliding environments as a singular assemblage constitutes the complex whole of an ecosystem. The relationship of an organism to its surroundings produces its field of subjectivity. Uexküll describes such a field as its *Umwelt*, a bubble surrounding each organism that is its own sole possession, and excludes all other organisms from the inside of that field (Buchanan, 2008, pp. 22–23). This image of the bubble suggests a private sphere of activity and meaning because each *Umwelt* is idiosyncratic, it is therefore incomprehensible to all others, epistemically isolated. But an organism's field of subjectivity affects ontologically interacts with other organisms and the land to shape the entire ecosystem. This constitutive production of daily activity is territorialization.

Understanding an ecosystem as the constitution and integration of territories merges ecology and ethology into a science that Andrea Brighenti calls territorology. All across lands and populations, organisms make marks that communicate how important they are wherever those marks can be seen. Marks make an organism exist somewhere, a presence to be reckoned with despite its physical absence. Marking constitutes territory, signs through which the marker exercises power across more space and time than its immediate physical presence (Brighenti, 2010, pp. 55–58). Marking territory suggests that you possess a private estate regarding which neighbors risk danger if they trespass. The possessor must manage the comings and goings through its territorial estate. A commonly recognized example of this marking behavior in ethology is birdsong, where individual birds produce unique song patterns. Their vocal performances establish dominance over a place, variously constraining the movement possibilities of potential competitors, mates, and predators wherever the songs can be heard. Wren songs warn intruders at high volumes and attract mates at lower intensities. Stagemaker birds hide and display bright plumage to complement their songs. Chaffinches perform a complex interplay of songs and subsongs, articulating a multifaceted personality of changing moods and reactions to wide ranges of circumstances (Deleuze and Guattari, 1987, pp. 323–333).

The concept of territory better accounts for how fields of subjectivity affect each other than Uexküll's original image of the bubble. The concept accounts for the ubiquity of semiotic interactions in ecosystems: the mark that defines their territory spreads their own field of subjectivity affects into their environment. Constituting territory through marks lets all such organisms affect the world beyond the limits of their field of perceptual affects (Guattari, 1996, pp. 164–165). Yet the concept of territory has its limitations. Understanding an ecosystem as a plurality of overlapping territories conceives it only in terms of its organisms, not the nonliving bodies like the soil or atmospheric gases. To understand an ecosystem comprehensively, we must consider territories a single class of constitutive affectivity among many in a gestalt ontology.

SELF AND TERRITORY

Any organism able to produce a mark on its environment constitutes territory in doing so, including self-conscious bodies such as humans. An important moral implication of gestalt ontology is that humanity cannot privilege its position in its own perception of the world. Properties of a body are not possessions of a discrete, isolatable unity, but are constituted from the continually active relations among bodies. Relations constitute qualitative properties, whether or not those relations involve self-conscious perceivers (Diehm, 2006, p. 26). An ontology of constitutive relations is a gestalt ontology, of which Næss was a pioneer. But my ontology goes further than Næss to reconcile difficult questions that the relationality of properties raises. Næss discusses the relational constitution of properties to emphasize the integration of all bodies, one of his most important ideas for ontological, ethical, and political contexts. But when his writing emphasizes the intrinsic value of all bodies, Næss writes as if each body was a discrete entity that possessed properties without any reference to relationality.

His environmentalist moral philosophy is thus incompatible with his ontology, ethics, and politics.

Understanding bodies as processes continually generating themselves can reconcile a moral principle of valuing a body for its singularity with the ontological principle of its processual nature. A process is never complete; it maintains a stable structure not because it freezes, but because its constitutive processes continue. If a body's constitutive processes change, it will transform, sometimes to its destruction. But there is no true destruction, only dissolution that conditions a new creation as matter and forces change the pattern of their assembly.

You may wonder where the subject is in all this talk of fields and processes. Subjectivity as the Western tradition of philosophy has generally portrayed it seems to have disappeared in the continual flux of relations that constitute assemblages. Previous attempts to build a concept of the gestalt subject, the self-as-place, has seen the place overwhelming the self, whether understood as territory, perceptual field, or field of affectivity. The typical roles of self as the exclusive seat of deliberation, will, and free action are, in a gestalt ontology, one kind of motion among a vast and diverse multiplicity. A self is one kind of agency in a universe constituted from the activities of many agencies. We construct moral systems as guidelines for self-conscious creatures who live in societies. The foundation of a moral system is a set of principles that depend on a concept of self or subject as the instigator of morally relevant action. The most radical element of environmental moral philosophy is how it dislodges the self or subject from the centre of morally relevant action. A gestalt conception of action pluralizes agency across all assemblages; agency is the constitutive relation that builds and transforms assemblages of bodies, forces, and histories of generation. The final chapter describes what kind of subject can exist in an ontology of relations while retaining its moral power as the engine of political action.

CHAPTER 7

ECOLOGICAL SELFHOOD, ECOLOGICAL LIFE

THE PURPOSE OF PHILOSOPHICAL THINKING IS TO CREATE NEW directions for human thought and open people to follow those paths. Changing how you conceive of reality itself and your own place in it changes how you live day to day, incrementally transforming your identity. If you have ever made a serious change in your life priorities, you remember your life before this shift as if you were a different person. I take this notion profoundly literally. If the power to transform yourself through thought is philosophy's highest, then philosophy is primarily an ethical matter. Beyond the scale of the self, philosophy's power becomes political. As human thinking transforms the identities of enough people, the character of a society accordingly transforms. Such transformation is a political revolution, a new image of humanity coming to life. This final chapter examines how transformations in thinking can lead to transformations in society, and the slow political revolution that ecological philosophy offers. As we change ourselves, so we change our civilization.

SELF IN THE WORLD

The concept of self unites the four domains of philosophical discourse: ontology, epistemology, ethics, and morality. As such, self is an ultimate metaphysical concept, and so resists the constraints of technical philosophical definition. But if you want to build a new morality and popularize it, that morality needs a conception of self, even as self resists easy conceptualization. The moralities based on the intrinsic value of nature, or the intrinsic value of singular identity, are radical departures from the mainstream Western philosophical and moral traditions. If the problems of our world require a new morality, such as that of integration and interdependence, we must change how we conceive of ourselves so this morality is intuitive. A philosophy that does not gain public attention as a serious morality is a political failure; it will have no effect on the world and eventually passes away, a forgotten idiosyncrasy.

You cannot pursue to the fullest the essential ethical question of what a person is, if you consider the inquiry to have no consequence for your own self-conception (Appiah, 2008, pp. 175–176). As you change how you conceive yourself, you transform the phenomenological structure of your own experience. Fear of radical change in our self-conception and the entire way we understand the world haunts us because of their profound risk to our identities. Such trepidation makes a metaphysical movement connecting an ethical question to any other philosophical domain especially sensitive to misstep. Næss's error in developing his concept of self-as-place was to move too quickly in *Ecology, Community, and Lifestyle* from sketching its ontology to stating its normative moral implications, principles for organizing an ecologically harmonious society.[1] A philosophy's movement from ontological matters to moral thinking requires a detailed account of the self because normative concepts are the social expression of how we each understand ourselves. Some approaches to philosophical thinking foreground concern for the self, particularly the Aristotelian-inspired traditions of virtue ethics and care for the self. Care and concern for the self is often taken up as the cultivation of virtue, an aesthetic exercise where your self is the object of artisanal activity (Swanton, 2003, p. 171). But an artisanal image of self-care is not appropriate to understand self as a place because it conceives body and mind as standing separate from the assembly of worldly processes. Ecosystemic processes constitute selfhood, but virtue ethics stops at the boundaries and behaviors of the human body, so it does not bridge ethics and morality, philosophy of selfhood and arguments over the normative principles on which society functions.

Environmental philosophy, even its ontological inquiries, inherited its moral goal from its genesis as the theoretical wing of a political movement. But a direct political articulation of the ontological concept of an ecosystem as a harmonious whole to which all parts submit can easily result in a politics of ecofascism, where the needs and desires of one component are unimportant and may be dismissed in calculating the good of the whole (Zimmerman, 1995, pp. 227–230). Yet dismissing a holist conception of care as ecofascism ignores the concept's subtle aspects. Specifically, understanding the self according to a holist ontology changes the concept of self in a manner that does challenge the claim that democracy is the universally superior political system. A holist concept of self-as-place identifies your ego with the entire assemblage of generative processes which produced you, and through which you produce others. You can participate and generate new organisms intentionally, as when a family plans to conceive a child. Phenomenological analysis shows that the intentional processes that generate self-consciousness also generate affects that constitute the subject's environment. But most processes that generate yourself and your ecosystem involve no intentionality and most self-conscious creatures rarely notice them. Phenomenology analyzes reflexive worldly consciousness and intentional perception, so has limited application to ecological thinking. Starting your philosophical investigation from intentionality makes subjective consciousness the locus and generator of action, and a dualism of self and world becomes inevitable. Such philosophical approaches as Husserl's and Merleau-Ponty's cannot conceive the subject as one body among many, an element in many overlapping assemblages of affects (Bello, 2009, pp. 62–65).

But phenomenology can offer important ecological insights. Consider the syllabically excessive subdiscipline of ecophenomenology. It tries valiantly to escape intentionality's gravity in phenomenology by building techniques to open an egoistic epistemic perspective to experience the genuinely different. Genuine difference is the inexhaustible contingency emerging from those affects that are present around a subject, but which its perceptual capacities do not condition, and resist total enclosure in the subject's field of affects, or *Umwelt* (Toadvine, 2010, pp. 346–348). Yet despite ecophenomenological philosophers' best intentions, this philosophy inherits its traditional parent's subject-centricity. An ecophenomenological subject retains its ontological privilege as the I, which holds it separate from environmental fields. Such separation enables an oppositional relationship to all that is not I. The I cannot be considered part of the greater diversity of the universe because phenomenological philosophy can only conceive of phenomena as existing in relation to the subject.

Examining ontological questions from a phenomenological perspective risks according the subject ontological priority over all else, even the conditions of its own existence. In phenomenological philosophy, you can build accounts of any object that a subject encounters, which completely and exhaustively define that object and all its possibilities. In all the variants of phenomenological philosophy, subjects alone are uniquely inexhaustible and therefore ultimately mysterious (Merleau-Ponty, 1962, p. 419). Considered on their own, phenomenological concepts leave me with an inverted Roquintin. Roquintin was alienated from the world by virtue of its mysterious singularity, the inexhaustibility of any account of an object using human thought's universally generalizable words. The phenomenological subject is the only mysterious singularity in a world that can be exhaustively defined. Alienation from nature runs to the deepest essence of humanity, a condition for the act of perception itself (Toadvine, 2010, p. 349).

Preserving and expanding individual humans' liberty is not a top priority for overcoming such alienation. Environmentalist politics focuses on safeguarding the productive dynamics of biospherical and ecosystemic processes from enormous industry. When this is a primary social and political goal, individual rights become controversial, and definitely far from inviolate. This shift in political priorities motivates James Lovelock's frustration with democracy (Hickman, 2010). Ecocentric moralities give a self no special moral and political consideration. So the most important strength of ecophenomenology is to maintain a moral centrality for the self in an ecocentric philosophy. While it may not succeed on its own in teaching people to perceive the singularity of worldly objects, it does repair the political alienation of a self's anxiety when facing the hard compromises of ecocentric morality.

Here is the metaphysical transition from ecological ontology to a democratic ecocentric morality. Ecophenomenology is the philosophical science that describes how a self can experience itself as a place, where its own subjectivity has equal ontological priority with all bodies and processes in ecosystemic fields. Self becomes one element among many in a wild flux of affectivity that includes its own generation, the conditions of that generation, its perception, and all the processes in its environment. The inescapable presence of self-consciousness in this flux gives it an indelible moral significance. The vocabulary of affectivity, despite its ontological framework, carries implicit moral implications when we discuss any subjectivity affects that generate phenomenological consciousness. All processes in the universe are ultimately integrated throughout their histories. So once at least one self-conscious subjectivity exists, its affects begin to spread moral significance throughout all processes because their fluctuations can affect moral agents.

The phenomenological element of ecophenomenology transforms self-as-place from an abstract ontological and ethical concept into an intentionally chosen consciousness of myself-as-place. Given this self-conception, the longest view of your constitutive relation will extend over the entire universe. Even so, the individual self-conscious subject does not disappear within this massive network, as if you were insignificant. We are small, but with a peculiar singularity in our power to think. Humans are, so far, the only known organisms having such high complexity of thought as self-consciousness and social personality. Minimal selfhood is a physical structure whose generative processes have a unique history, but moral reasoning requires a more complex selfhood than autopoiesis alone.

Because ecological philosophy is process thinking, ecophenomenology must accordingly account for the temporal finitude of all processes. Everything is mortal. But phenomenologies deal not in processes, but presences. A particular presence is mortal only in reference to the subject itself. As Ted Toadvine writes, "The experience of the thing is always an *in-itself-for-me*" (Toadvine, 2010, p. 348). Not only does ecophenomenology conceive bodies and processes as presences, but also presence is a function of the subject's experience. A process could have begun long before intersecting with a field of perceptual affects to become present in experience. And a process can continue without dissolution long after passing away from that field. Because only self-conscious processes can reason, reasoning is possible exclusively within this field of self-attention. Moral significance becomes ubiquitous as subjectivities interact with the world, but moral subjectivity itself requires self-consciousness. A philosophy of affectivity understands experience as a body's perceptual apprehension of its field of perceptual affects. A person always experiences phenomenological selfhood in this minimal sense.

The Paradox of the Prepersonal Self

Imagining a world without yourself or without humanity conjures an image that is not only alien, but from which we feel alienated. This alienation is not merely an intuitive strangeness we may feel when reading descriptions of a dilapidated 7th Avenue whose signs are never illuminated, where pavement is cracked with long, jagged invasions of weeds growing without restraint over long-ago-flooded subway tunnels (Weisman, 2008, pp. 25–34). In all the experience of a self-conscious subject, its place, and its history, there is always a minimal phenomenological selfhood; the prepersonal or a-personal has never truly been present in experience. There is no possible arrangement of a perceptual apparatus to perceive any kind of prepersonal or a-personal. The only way to conceive of the prepersonal is through abstract reasoning (Merleau-Ponty, 1962, pp. 241–242).

The prepersonal is how Merleau-Ponty understood the invisible conditions for the appearance of the perceptible. We cannot make sense of a world without this minimal self-consciousness, a world without thought, because the act of making sense is itself a matter of thinking. We can lay out propositions that describe a world without self-conscious perceivers, but we can only make sense of it as a self-conscious perceiver (Barbaras, 2001, p. 29). Any attempt to understand a world without perceptual fields will always involve at least the perceptual field of the one who understands it. For this reason, some commentators on Merleau-Ponty's work take his ideas to imply an epistemic solipsism that grounds an ontological one, in which the entire universe and its history is a function of subjective activity.

If we accept this, then we must conclude that no subject can ever know the world itself, because all knowing refers back to the self-conscious body. As the locus of all experienced activity, a subject can have no knowledge of nature that avoids conditioning by the structure of its field of perceptual affects (Wolfe, 2009, p. 220). Our perception even conditions abstract propositions that describe such simple facts as the atomic composition of compounds because we only learn them through self-conscious perception. To articulate the problem in terms of autopoiesis and systems theory, because the subject can access only those meanings that it generates itself, we cannot understand other bodies without subjective epistemic conditioning (Bello, 2009, p. 70). Yet there are facts of nature that exist without direct relation to a subject, regardless of its existence (Barbaras, 2001, p. 27).

Interpreting phenomenology to imply that a subject is necessary for the existence of the universe derives ontological scope from its epistemic conclusion. Paradoxes emerge from all such movements of thought between philosophical domains, but this case's prima facie contradictions provoke a new way of thinking. This interpretation of phenomenology is humble in accepting the limitations of the subject's scale and epistemic powers, yet simultaneously commits the philosophical hubris of conceiving it as the source of any coherence in the universe. A paradox with a similar structure occurs in Næss's writing when he empowers the ego of the environmentalist activist to build a society whose morals and ethics are defined by humility before nature. The universe is incomprehensible without the striving of self-conscious bodies like humans to comprehend it because the facts of nature cannot be known without knowers. Yet only physical movement itself is sufficient to constitute orders of being. Ontological matters are the conditions of epistemic matters. The possibility of epistemic processes only exists once there are epistemic agents, as the first autopoietic structures emerge. Ethical and moral processes become possible only when organisms develop the ability to reflect on their experiences. Human personalities and societies run concurrently with the ontological and epistemic processes that have continued for billions of years (Deleuze, 1993, pp. 5–8).

Autopoietic structure is an ontological condition for the possibility of a self, because only an autopoietic body among all carbon-based assemblages of molecules can generate perceptual affects. Autopoiesis is a sufficient condition the most basic epistemic activity of directed motion, the simplest individuality. "Individuality in this case corresponds to a *formal self-identity*—to an invariant dynamic pattern that is produced, maintained, and realized by the system itself" (Thompson, 2007, p. 75, emphasis added). But the self is more than a body standing out from a background, heterogeneity emerging from the homogeneous. Ontological individuality does not include the epistemic powers of a self, let alone its ethical and moral powers. Yet another reason why Evan Thompson considered autopoiesis as an actual self, beyond its foundational activity of sense making, was that he considered the autopoietic body a prereflective self, explicitly borrowing the concept from Merleau-Ponty. Thompson writes, "Pre-reflective experience is logically prior to reflection, for reflection presupposes something to reflect upon; and it is temporally prior to reflection, for what one reflects upon is a hitherto unreflected experience" (Thompson, 2007, p. 250). This is the problem of self-production applied to consciousness.

Contrary to that argument, self-consciousness requires no prereflective consciousness to exist in an individual to activate reflection. A self-conscious personality begins when a perceiving body reflects on itself for the first time. Each person's self-consciousness emerges at a unique time, so there is some small section in each of our individual pasts when we were not and have never been aware of ourselves. While a body does carry out that act of reflection, you the organism are not immediately transformed into a new order of being as self. The first stirrings of an I are subtle and tentative moments in which you are aware of your experience as what is happening to you; your first primitive thoughts as a subject treat you as an object. But a prereflective consciousness is the fulcrum of Merleau-Ponty's invisible, a shadow of our self-awareness whose nature makes it inevitably imperceptible. You cannot perceive yourself free of reflectivity, because perceiving yourself is this reflectivity. The prereflective consciousness must be a different entity from you as a self-conscious body because of the paradox of self-production. A self cannot produce itself because doing so would require it to exist prior to its production. But the paradox of self-production is only valid for essentially static entities, and there are no such things. All is process, including you and I. Self-production is literally a process that constitutes a self or selves, and there need be no proto-self prior to the existence of an actual self to build it. Nonintentional processes like metabolic chemical reactions can constitute intentional bodies. A self emerges from a perceptual field that notices its own activity.

SELF AS DYNAMISM AND CREATIVE DYNAMO

We typically consider a self as a static, discrete body, but it is actually a dynamic process. Each act of reflective thought or perception discovers a body different from the previous reflective act. Time has passed: Our field of perceptual affects has changed. When we conceive ourselves as static bodies, our mistake blocks us from articulating our powers to the fullest. If I can be cheeky in my description, an ecological philosophy must conceive of the subject without subjecthood. Gilles Deleuze and Félix Guattari lead the way in overcoming the concept of the subject as a unified stability. Their alternative is a vision of subjectivity as a process, a dynamic multiplicity from its generation throughout its continuing existence. As a process, subjectivity is always open to change, so if a subject is static then there can be only one kind. That kind would be universal and necessary. Deleuze and Guattari's *Anti-Oedipus* explores the political implications of this ontological topic. They analyze how any attempt to build a universal and necessary concept of subjectivity—whether in philosophy, psychology, or everyday morality—is a means of oppression (Deleuze and Guattari, 1983, xi–xiv).

Beyond diagnosis, Deleuze and Guattari's work devises ways of thinking that escape this oppression that imposes necessary identity and subjecthood. Their emancipatory purpose forces them to assume the following: (1) homogeneity or conformity is a needlessly imposed unity that harms the freedom of one's personality; therefore (2) we free ourselves by a refusal to accept unity of any kind and create a new trajectory for our lives. Although you should best approach any kind of conformity with a critical eye, it is not necessary to fear unity per se. The extremity of their language bludgeons presumptions and perspectives across the philosophical disciplines that biased practitioners against process thinking. Self-identified process philosophy lies outside the Western tradition's mainstream, and Deleuze shaped his historical influences as a self-conscious countertradition. Guattari began his career as a discipline of Jacques Lacan's Freudian tradition, which treated a compendium of symbols from case readings as the universal structure of the mind. Lacan was personally offended by Guattari's professional break from his school in *Anti-Oedipus* (Dosse, 2010, pp. 184–185).

But an actual object that we call a unity does not have a permanently unchanging nature. Bodies are assemblages of processes and naive unities are not real. A functional unity preserves the multiplicity of its constitutive processes, as in Aldo Leopold's example of the many creatures living in the husk of a dead tree. But some processes force a destructive, homogenizing unity on its constituents, suppressing multiplicities to totalize an ecosystem according to a single, exclusive program as in Leopold's example of the monoculture farm. The impulse to liberate yourself from conformist surroundings is the tendency for a complex ecosystem to diversify. Conceiving subjectivity as a process finds the essential vision of Enlightenment liberal freedom as it occurs in the human subject occurring in ecosystems themselves. The politics of singular freedom resisting the uniformity of authority is the politics of both the subject and the ecosystem. Such a political perspective can also guide the activist in her daily life in how best to change her society to adopt an ecocentric morality. The foundation lies in what a subjectivity's field of affects requires to break from a destructive, homogenizing, regime of uniformity to become a unity that encourages its own constituent processes. A self-sustaining process should also be multifaceted, diverse, and flexible so that its range of action can adjust to environmental changes that are genuinely beyond its control.

Living an ecocentric lifestyle must start with conceiving yourself as a process that can sustain itself over a finite duration. The static subject is a unity that reduces all its potential perceptual affects to an internally consistent set of possibilities: the I. Self-consciousness, in considering itself a unified I, prioritizes consistency over internal diversity and flexibility. Becoming consistent homogenizes the transformative dynamism of perceptual affects. In considering yourself a unified I, you suppress the dynamism of your own body, forcing yourself to conform to a single vision of what humanity is. Deleuze and Guattari call this homogenizing conception of self the phenomenon of faciality, evoking the surreal imagery of minimalist masks whose faces are no more than three black holes. Personal and social norms endorsing consistency and authenticity of identity ignore and suppress the internal diversity of our personalities (Deleuze and Guattari, 1987, pp. 170–171).[2]

However, the practical matters of daily survival require maintaining your identity's stability. You must turn away from the entire flux of perceptual affects to focus on a single desire immediately important to sustain you. In this moment of life, you understand yourself as a static personality. In times of danger and chaos, you can act only after concentrating on a calm center in your thinking. You might think that because conceiving of itself as a homogeneously unified I totalizes a body, it is easy to extricate yourself from such thinking: "I can stop anytime I want!" But the postulate of a homogeneous I is more powerful an addiction than any mere drug, because we can slide too easily from a desire to maintain the minimal stability for any sane life, into a reactive fear of change that is unhealthy both for subjectivities and ecosystems. But the experience of focused homogeneity is ubiquitous, because any intentional action requires a stable focus of perception or thought. If you take any stability to be essentially restrictive or oppressive, and also an inescapable aspect of one's existence, then all you can offer are promises of escape through transcending existence. Empty promises (Deleuze and Guattari, 1987, p. 186).

Of course, an assemblage will break apart if it wildly fluctuates at enormous intensities. It must have some minimal stability to avoid catastrophic collapse. Destabilize a system too quickly or without sufficient care and all flies to pieces (Deleuze and Guattari, 1987, p. 161). Yet an absolutely stable assemblage that does not change or move at all is dead. Even the constant stability of an autopoietic body's boundary cannot exist without metabolic chemical reactions continually roiling underneath it (Maturana and Varela, 1980, pp. 78–80). A totalizing self-identity is worth escaping because embracing homogeneity in your identity requires conceiving of yourself as an unchangeable unity, where all you can be is what you have been and now are. The current crisis of humanity's enormous industry teaches us that reality manifests problems that cannot be solved with the ideas and philosophies of the past. A self-identity that maintains its homogeneity at all costs cannot adapt to novelty. Valorizing the stability of your identity above all other concerns suppresses any dynamisms among its fields of perceptual or cognitive affects. Your own tendency to unify and homogenize shuts down any divergence from your identity's existing parameters. Such a personality cannot even tolerate diversity in its environment, remaking the world in its image. "It propagates waves of sameness until those who resist identification have been wiped out" (Deleuze and Guattari, 1987, p. 178).

Resisting homogenizing social and environmental forces is an ethical stance, a way of conducting yourself that informs every feature of your self-conscious identity. Such processes of homogenization and divergence occur among individual organisms, or proliferating across an entire society's economic, political, or ecological order (DeLanda, 2000, pp. 272–273). One way to articulate this ethical stance against homogeneity is to create a moral system whose norms encourage actions that increase and safeguard diversity. Safeguarding diversity is a consensus normative principle in environmental moral philosophy. Disagreement arises across the discipline not over the principle, but how best to act on it. Different approaches include respect for the nonhuman subject matter of scientific inquiry (Jickling and Paquet, 2005, pp. 124–125), experimenting with the different ways cultural traditions regard nature and the nonhuman (Abram, 1996, p. 10), or using emotional responses of wonder at the diversity of life to justify moral commands (Moore, 2005, p. 277). Arne Næss laid out his own approach in the guidelines of Ecosophy T (Næss, 2005, vol. 10, p. 535). Building such normative systems that value diversity are experiments to discover what kinds of thinking can overcome the totalizing powers of self-consciousness.

Opening your self-consciousness to understanding difference requires overcoming our all-too-common fears of change to embrace the heterogeneity of existence in divergence. Most important is understanding that divergence is not inherently destructive. The development of new capacities can be a productive divergence from a formerly stable identity, as an individual breaks with old habits of living that are no longer appropriate for her life. Even such simple acts as learning to drive, to swim, or to speak a foreign language are experiments in identity creation (DeLanda, 2006, p. 50). Human powers of self-consciousness actually make such breaks easier because we can control our transformations. Experiments in breaking down and reconstituting yourself can be more easily botched the further they depart from your usual actions.

Deleuze and Guattari discuss different ways that a subject can experiment with the dynamism of affectivity to disintegrate her totalizing field of selfhood. They aim to spark creative inspiration instead of bland, cookie-cutter imitation because a new subjectivity can only succeed when its creation accounts for its singular environment and needs (Bogue, 2009, p. 46). An individual in her own singular situation may require a completely different process, but these examples are not meant as step-by-step guidelines, only inspirations and demonstrations that destroying and rebuilding your own subjectivity is possible. One example in their collaborative work is a sexual masochism that forcibly seals all entrance and exit points for the body, and covers or blocks all perceptual organs except touch. Another is an all-encompassing heroin addiction and opiate coma, as described in the work of William Burroughs, that totally freezes the subject, annihilating his personality (Deleuze and Guattari, 1987, pp. 151–154).

Guattari discovered other examples of how to disintegrate the subject in his work as a doctor at La Borde clinic. Treating the severely mentally ill helped Guattari understand schizophrenic behavior as an escape from all constraints of identity and society. His patients give in to forces that fragment the subject at amazing degrees of intensity (Guattari, 1996, pp. 39–40).[3] "The forces of attraction and repulsion, of soaring ascents and plunging falls, produce a series of intensive states . . . The subject spreads itself out along the entire circumference of the circle, the centre of which has been abandoned by the ego" (Deleuze and Guattari, 1983, p. 21). Such processes are easy to botch because a field of subjectivity affects is extremely fragile at extreme intensities of fragmentation and stillness. An organism destroys its subjectivity's totalizing element, but also its entire field of perceptual affects. The high possibility of disaster would appear to justify the fear of any departure from homogeneity. But as Deleuze, Guattari, and DeLanda stress, attentive care and control of the transformative process is all that we need to avoid permanently shattering ourselves.

For all its danger, breakdown is the easiest part of transforming your subjectivity. We may purge our old homogenizing forces, but the subjectivizing process will dissipate if it does not constitute some new unity. This new identity must not make the old mistakes of conformity and homogenization for their own sake. Putting a new face on the same problems is just another way to botch the change, putting yourself through a dangerous process for nothing. Having removed your personality's destructively totalizing elements, you must rebuild a unity that includes diversity, a genuine gestalt body. *A Thousand Plateaus* calls this creative regeneration of subjectivity becoming-animal. The term sounds as if we become like an animal, but this is not adequate to what Deleuze and Guattari ask it to do. Their examples are of people abandoning conformity to human society for what appear to be imitations of the ways of animals.

They begin with an illustration from the film *Willard*, in which the protagonist alienates himself from human society and lives according to the social structures of rats (Deleuze and Guattari, 1987, p. 233). But Willard does not literally live *like* a rat, scurrying on the floor and in walls. Becoming like an animal would simply trade a human mode of totalizing subjectivity for some other species' totalizing subjectivity. Both smooth diversity into a single way of life. Imitation cannot escape from totalization into heterogeneity (Deleuze and Guattari, 1986, pp. 39–40). An assemblage has a greater range of capacities than it can ever articulate in its lifetime. These capacities include all articulations that its internal processes can generate, all interactions with every other body with which it could ever come into contact, and all the results of those worldly interactions. Topological mathematics can describe these limit points of possible motion as a quantifiable phase space for a process' movement (DeLanda, 2002, p. 80). Someone open to the heterogeneity of her identity may, realizing that these possibilities are available, consider changing herself. If you believe that divergence from your identity is impossible, you will never achieve it.

Openness in thought to change yourself is sufficient to overcome, at least negatively, your subjectivity's totalizing identity. A totalizing identity is unable to conceive or even seriously imagine most of its own possible movements, and if he does notice them, he will convince himself that those movements are impossible. Maintaining that inconceivability is what totalizing identity does. A subject's social and ecosystemic environment can also reinforce the totalization of your identity and prevent your divergence from established homogeneities. Many of these impediments cannot be overcome. I cannot teach myself to fly, or raze an urban industrial facility to build a nature reserve by myself. But social and political activism can overcome a lot. You may have grown up in a political culture that is hostile to environmentalism, and encountered genuine dissenting perspectives so rarely that you do not know how to take them seriously. Perhaps you live in Saudi Arabia, where the state has gained so much economic profit from the petroleum industry and subsidized oil products so heavily that the notion of its scarcity is literally inconceivable. It is easy to say that opening your thought to new possibilities is all you need, but actually doing so can be extremely difficult.

Concepts like Deleuze and Guattari's becoming-animal offer tools for a kind of philosophical therapy to ease your transition to multifaceted modes of thinking and living. Rebelling against a totalizing identity is not enough. Rebellion and subversion are entirely reactive because they always need a target to rebel against or subvert. Defining yourself as a rebel sends you looking for an authority to rebel against, and you remain as dependent on that authority as its most dutiful servant (Guattari, 1996, pp. 62–63). Destabilizing an identity must be the beginning of a longer, more complex, dangerous process to constitute a new identity (Kazarian, 1998, p. 103). Their philosophy aims to convince readers that a change of identity that you may think impossible is not only possible but worth embracing as your liberation. Guattari writes, "My question therefore is, can man become the founder of his own law?" (Guattari, 1996, p. 66). He asks if a person can create a new normality, knowing that our old normality need not constrain us.

The Nature of Identity

Only our activity constitutes our identity. We usually change our identities for practical reasons: the world has changed such that the old normality is counterproductive. Such a change may run against the moral systems of your community, which is why your peers may shame you. But in the *absolute conception* of an act's singularity, its existence needs no more than that it is done (Sartre, 2007, pp. 24–25). Whether the consequences are ultimately beneficial or destructive to the actor, her community, and her world is an important, but separate, matter. An act is justified by the *practical difference* it makes for an organism, community, species, or ecosystem. Any thoroughly ecological ethical perspective and whatever moral systems we develop within it would distinguish these two conceptions of value. The absolute conception holds a body or process to be valuable for the singularity of its existence, that there will never be another body precisely the same as this. The practical value of an object is how some body or process affects you or the systems on which you depend to live.

Value must have this duality: what is so harmful to me that my survival depends on its destruction also has value in its existence. There is value in even the greatest ecological disasters, a sublime disaster of such profound and massive idiocy as to become theatrical. Colossal follies are still worthy of respect. Transitioning to an ecocentric morality with a dual conception of value is simply the self-conscious decision to change habits and change how we understand the world. The worth of any new normality is determined only by whether such a style of action can survive and thrive. When humanity wields powers that can be destructive on the scale of enormous industry, ecocentric ethics have incredible survival value because they help us understand the genuine scale of the effects of our smallest actions. From the absolute conception of singularity, the Pacific Trash Vortex, for example, is absolutely valuable. As well, in a colloquial sense, it can be beneficial to the world, even though that benefit may be no more than inspiring a bleak comedy routine. Its destructive effects on the ecosystems of the Pacific Ocean utterly outweigh the moments of cynical amusement some humans may derive from it. Justification by singularity can only go so far in practical matters.

Changing identities is not a simple matter when it implies such a profound transformation of self-image. Deleuze and Guattari discuss becoming-animal, becoming-woman, becoming-child, becoming-molecular, and becoming-intensity. Each of these transformative movements escapes the totalization of subjecthood because you change your activities so that they are not bound by a pattern that demands conformity without diversity. A transformative exemplar is a guide that we require to change our lives, a guiding example that we should approach as if it were an attractor in the phase space of our capacities. We should never aim to duplicate our guide, because duplication is homogenizing imitation. Instead, incorporate your role model as the primary influence in your existence, a trailblazer of your new normality (Deleuze, and Guattari, 1987, p. 274).

Each arrangement of bodies, movements, and energy flows is absolutely singular, which means that we cannot distinguish them by the mere fact of their singularity. But there are two ways to measure singularity. We can develop a quantitative measure for each aspect of a body, whether a discrete property or the intensity, duration, or periodicity of a component process. We can design a huge graph that represents each aspect as a dimension. Such a graph can have thousands or even millions of dimensions, so it would only be sensible as a computer model, not as a visual aid. Each of our measurements would be one point on the relevant axis representing one aspect. "The sum total of the material elements belonging to it under given relations of movement and rest, speed and slowness (longitude); the sum total of the intensive affects it is capable of at a given power of degree of potential (latitude). Nothing but affects and local movements, differential speeds" (Deleuze and Guattari, 1987, p. 260). These measurements plot a body's capacities for movement and affectivity to their farthest range of possibility. Only such a comprehensive mapping is adequate to an assemblage's singularity (Deleuze and Guattari, 1987, p. 262). Genuinely adequate knowledge requires a sufficiently powerful computer to perform the mathematical calculations to render all the qualities of a body in a quantitative matrix. That account would be so incredibly long and complex, that today's most advanced computers are not yet up to the task.

Comprehensive knowledge of a particular body in itself still remains inadequate to the complete situation of its singularity, which requires comparing it with all its surrounding assemblages and processes at the moment and in the past. The degree of contrast measures that body's degree of singularity in its local context, the practical dimension of a body's singularity. We quantify the degree of a body's degree of departure from its previous state and difference from surrounding and related bodies. While I draw heavily from the precedent of Deleuze and Guattari in developing all aspects of my conception of singularity, their own work focuses on contrastive singularity because their political concern is escaping the totalization of a homogeneous identity. They are wary of using general propositions to describe the singularity of a body. Like Sartre's Roquentin, they know such language enfolds the particular into a generic definition or statement that can never completely capture the uniqueness of its identity. Roquintin despairs at natural language's inadequacy to existence, but Deleuze, Guattari, and I love the singular for how it inevitably slips away from comprehension within the limits of the general.

Generalities focus only on those aspects of a body that are shared among others, accounting for a body as an instantiation or token of a type. Understanding something in terms of how it fits into generic taxonomies has its uses, but ignores its singularity (Deleuze and Guattari, 1987, pp. 264–265). The best chance of escaping the totalization of a homogeneous identity lies in developing the greatest possible contrast of your own body's singularity from all other similarly totalized bodies. Nothing can sustain such rebellion indefinitely, but can serve the immediate task of escaping homogeneity. Humans, with our philosophical and computational powers, can map the singularity of bodies conceptually and mathematically. We can make such maps of ourselves to understand our own capacities for radical departure from our homogenized neighbors.

Deleuze and Guattari provide the last block in an ecological philosophy that achieves what Luc Ferry's arch-humanist critique claimed was impossible for environmentalism. The power to escape from conformity is essential for both traditional liberal politics and activist environmentalism. Our conceptions of what the world is condition what we consider essentially important to our lives in it, our values. Our values are the premises from which we reason through what is right or wrong, good or bad, to deal with each moral problem that the world presents us. Holist ecocentric morality, which debases the value of an individual to preserve a supposedly harmonious whole, is rightly rejected. Such holism is not grounded in the truths of our ecological relationships, but in the essentialist mythmaking of the absolute dichotomy. Thinking ecologically across all philosophical domains understands the absolute value and ubiquity of singularity at all scales of existence. All wholes assemble greater wholes, and each whole depends on all its parts just as much as each part depends on the whole.

Identifying Singularity

We cannot act on our ecocentric values of singularity without understanding how its absolute and practical measures relate in our worldly lives. Such understanding rests in two fundamental knowledge operations. Constitutive identification is the epistemic acknowledgment of a body's absolute singularity, understanding a body such that the knowledge provokes wonder at its uniqueness. Contrastive distinction measures practical differences, how a body diverges from the processes that surround it and that generate it. Contrastive distinction is also the starting point to map the empowering and destructive affects that a body can have on other bodies such as myself.

These two ways to measure singularity merge in the concept of *style*. If you want to teach someone what singularity is, art is probably the best example. A singularity's content and its manner of expression are inextricably bound (Merleau-Ponty, 1962, p. 174). A body's style is a whole constituted from the relations among all its parts, and cannot be considered separately from them. Any attempt to analyze a body by breaking it down into individual elements and relations will ignore the dynamism as all these parts work together as a process. Adequately understanding a body must include the processes constituted when all a body's parts function to assemble the whole. Merleau-Ponty describes an artistic example "As for the novel, although its plot can be summarized and the 'thought' of the writer lends itself to abstract expression…the novelist's task is not to expound ideas or even analyze characters, but to depict an inter-human event….A novel, poem, picture, or musical work are individuals," each having a particular style (Merleau-Ponty, 1962, p. 175). An artwork's greatest power is not the expression of a general idea, but the depiction of an event, the interaction of many disparate processes to constitute a complex whole.

He only gestures at this conception of an event, but Deleuze and Guattari describe it in detail as the sudden intersection of processes from multiple levels of analysis at a single place. When many different processes meet, their interrelations produce a complex whole that could never have been predicted from the analysis of each process' history in isolation (Deleuze and Guattari, 1987, pp. 202–203). Merleau-Ponty's description of artworks as individuals each having a peculiar style accords with the existentialist conclusion that general statements about a body can never be adequate to the peculiarities of its singular constitution. Singularity is not only a body's property. It is an event constituted by a body's own activity, the convergence and integration of all the processes that generate it (Deleuze and Guattari, 1987, p. 265). Art is not only an excellent illustration to teach people about singularity. An artwork can inspire a trajectory for a person's transformation because it can articulate an idea in the complexity of an event and our encounter with one. An artwork can achieve implicitly what a philosophical work can achieve explicitly: articulating a concept that can transform your thinking and your life along with it, spurring a singularizing event.

When we take up an artwork into our thinking, it produces an event that articulates itself as a process that transforms a subjectivity (Deleuze and Guattari, 1994, p. 21). As a subjectivity singularizes itself, it transforms its environment, because they are so intimately integrated. This is a moral matter, a question of what we should do. The moral aspects of singularization can repair the problems from which environmental philosophy tends to suffer thanks to its origin in political movements. Progress in any field of human endeavor relies on change and stability. Any human endeavor is a tradition of social activity that maintains a minimal continuity over time, sometimes multiple generations. But a tradition's practitioners must also be critical enough of their habits and history to recognize when old ways of thinking, even those central to a tradition that has long held social value, become inadequate to the world because new pressures have rendered it literally obsolete. Singularization is not only a break from continuity with your immediate surroundings, but from a path of historical development. A work of philosophy that continues an established tradition without reflective critical thought homogenizes itself, as it focuses only on conserving its continuity with those texts that have influenced it (Ford, 2005, p. 42). In departing from conservative standards of adherence to an obsolete tradition, the work becomes a new model for future writers and works to follow (Deleuze and Guattari, 1994, pp. 67–70).[4]

An experiment in thinking is creative when it develops a capacity in thought that has never been explored, or at least that the current mainstreams have ignored. Such an experiment is public because a work of philosophy is a communication, whether a book, article, essay, or blog. It can be read and discussed, spreading its influence through written and spoken media (Kazarian, 1998, p. 102). One way scholars trace the history of philosophy is by tracking influence and discussion. Works that destabilized commonly accepted traditions of thinking, encouraged departure from those traditions, and inspired new traditions are historically significant (Ford, 2005, pp. 53–56). Environmental politics, activism, and the ideas that grew from them constitute environmental philosophy's departure.

The new tradition of environmental philosophy is a public experiment in thinking consisting not only an innovative ontology, but also self-conception, scientific practice, as well as social and political arrangements, institutions, and norms. Underlying environmental philosophy's experiment in transforming individuals, societies, industries, and ecologies is the ontological concept that relations are the primary constituents of bodies. Relations constitute fields of affectivity that integrate many processes into complex wholes. Fields sometimes integrate in patterns that maintain a process' stability, while some integrate in a manner that destroys stability. When a metabolic chemical process constitutes an autopoietic body, that body begins producing a field of perceptual affects. When that field of perceptual affects constitutes reflective processes, it becomes self-conscious. Deleuze and Guattari describe how the reflective processes of self-consciousness tend to totalize a subjectivity, and how an individual can escape this totalization to explore different possibilities for action (Deleuze and Guattari, 1987, p. 291). "Life is a work in progress with no goal in sight, only the tireless endeavour to explore new possibilities" (Guattari, 2008, p. 8).

Singularization: Revolution Begins with One Person

Guattari's solo writings explicitly apply his and Deleuze's lessons on destabilizing subjecthood to the political problems of environmental philosophy. Guattari understands the fields of affectivity that constitute Earth as three parallel ecologies: the mental, or individual subjectivity; social assemblages, including political and civic institutions; and the constitution and interaction of ecosystems across Earth's biosphere (Guattari, 2008, pp. 19–20). These three types of organically generated fields should not be arranged in a hierarchy. A society is assembled from many individuals, but the micro does not determine the macro, or vice versa. Ecosystemic and climatic conditions limit a society's and individual's possibilities for movement. You cannot grow coffee in the Baltics, or drill for oil year-round at Mount Erebus in Antarctica. Yet while the ecosystemic conditions social development and individual thought, such a function does not make the ecosystemic interests necessarily more important in all practical problems.

A process's identity does not change when it crosses multiple ecologies, only the context and setting of its appearance. Changes of scale, in contrast, can transform the character of a process ontologically. A current of water in a small tank will move very differently than a current of water in motion from north Atlantic to south Atlantic. A river can flow through a forest, by a village in that forest, and around a swimmer from that village without seriously changing. The purpose for which we want to understand a phenomenon is not our only guide to its best possible explanation. In terms of the ontological question of what is happening, the best possible explanation depends on which venue of affectivity primarily generates a phenomenon.[5] Understanding the world requires describing, sorting, comparing, and contrasting the three venues of affectivity to analyze properly all the multifaceted interactions among complex wholes. Distinguishing the three venues by parallelism keeps phenomena congruent across their contexts (Deleuze, 1990, p. 109). Activity generated in one of the three venues constitutes complex affects there and in the other two. Very different bodies and fields constitute each venue, but their mutual affectivity intimately connects them. We can clean up pollution, but we cannot repair an ecology beyond such superficial maintenance without repairing in parallel our own subjectivities. We individuals must articulate a sustainable and flexible pattern of thinking to build a sustainable and flexible society. We individuals simultaneously articulate the society that our activity constitutes into a sustainable and flexible technology with global scope (Conley, 2009, pp. 116–128).

Producing a new way of life begins when a single person breaks down a totalizing identity that conforms with common morals for conformity's sake alone. If this person becomes others' guide for action, she inspires others to produce similar new ways of life, until we have a social movement that affects a globalized humanity, and controls technologies powerful enough to transform ecosystems on a planetary scale (Guattari, 2008, pp. 23–27). A global revolution to constitute an ecocentric moral and political society might begin when an individual resingularizes herself as a place. Unlike more problematic ecocentrisms, such a transformation of social morality and global ecosystemic order preserves the importance of the individual. A transformation of individual morality begins a process of planetary transformation (Starkey, 2007, pp. 154–157). Democracy at its most intense is large-scale aggregate liberation: individual subjectivities simultaneously transform, constituting a social movement. Singularization processes break down a normality to constitute a new normal, idiosyncratic at first. The most radical transformation into a new individual, social, and ecosystemic assemblage always retains just enough of the old order to remain stable. Yet its new context utterly transforms this old element (Guattari, 2008, pp. 30–32).

In all this interdependent complexity of a global transformation, we as individuals can still control our subjectivity's power and role in it. You can modify your self-consciousness, your habits of thought and action, to recreate your subjectivity. Such a subject transforms society simply by moving in it, inspiring change in others merely through interacting with them. With sufficient power, such a society can constitute a new ecosystemic arrangement on Earth, a literally changed planet. The three venues of individual thought, social institutions, and ecosystemic assemblages are delineated as they are because that tripartite distinction is the clearest map of what kinds of factors must collide for a singularization to happen.

These are all contingent transformations. A society just happens to discover that it has developed in an environment offering the freest conditions for individual thought that escapes the status quo. Deleuze and Guattari discuss why the particular philosophical tradition of the West arose as it did, differently than in China, India, or somewhere else: the conditions of pre-Periclean Greece just happened to be right for this specific kind of thinking (Rorty, 1989, p. 61).[6] But we can ask the same question—Why here rather than there?—of any process (Deleuze and Guattari, 1994, pp. 95–97). A philosophical concept is a thought that can singularize an individual, society, or ecosystem, turning it away from long-established patterns of development into a different direction. A concept becomes able to singularize reality when an individual thinker connects it with social and environmental milieux such that bodies in all these contexts change. Thinking is at its most powerful when it is already operating in the world, engaged in changing status quo arrangements (Deleuze and Guattari, 1994, p. 92). Where conditions are not amenable to constituting fields of affects that can foster this engaged, revolutionary thought, revolution does not happen. Our physical and social environments are where revolutionary thinking plays out, but they only set conditions.

Revolutionary thinking begins with a thinker because our self-conscious affectivity can experiment privately in thought. Private experimentation does not carry the risks of real-world social, ecosystemic, and ideological experiments. Deleuze and Guattari mention the American and Russian revolutions as examples (Deleuze and Guattari, 1994, p. 100). Such dangers fundamentally challenge any philosophy that begins as a politics. Utopian political revolutions are movements to transform a society along a set of ideals for perfection. The visions of ecosystemically stable, perfectly sustainable communities that writers like Arne Næss abstractly describe are utopian in this sense. Utopian revolutions promise so much, but never in their actual history do they deliver (Deleuze and Guattari, 1994, p. 103). Yet the goal of environmentalist politics is a successful social revolution, and had been so before the philosophical community even codified it as such (Light, 1996, p. 327).

Ethics is the philosophical domain of subjectivity and its limits. One of those limits is that a single self-conscious organism cannot transform a society through force of will alone because the world is too huge and complicated. A process can easily botch itself if it crosses the tripartite venues of thought, society, and world without the proper power at the proper intensity. This power lies not in brute force, but in inclusion, incorporating an individual's own practical tendencies into the larger transformative movement. We can only achieve this by working together with other people. A self-conscious body develops all the conceptual and linguistic tools to become a genuine environmental philosopher through participating with others to find ecologically sustainable lifestyles, building scientific systems to measure and understand perceptual, technological, and ecosystemic affects, and learning the strengths and weaknesses of theatrical enormous industry. Philosophy is a posthuman discipline, the practice of thought to overcome the current limits of humanity's powers, so an environmental philosopher works to overcome humanity's ecologically destructive thoughts, powers, and lifestyles to live and think sustainably.

The Art of Becoming

Deleuze and Guattari discuss several figures from literature to illustrate becoming-animal and becoming-woman: the rat-man-protagonist of *Willard*; the wolf-men of Sigmund Freud and Jorge Luis Borges; Herman Melville's Ahab, mapping his life onto a whale's; H. P. Lovecraft's outsiders; the gender fluidity of Virginia Woolf's Orlando; and Marcel Proust's reimagining of Albert as Albertine. These examples come from art because ethics, the philosophical domain dealing with how to constitute subjectivity, is fundamentally aesthetic. As Merleau-Ponty said, the self is a work of art, and the philosophy of subjectivity explores and develops techniques and concepts for crafting a self. Guattari understands aesthetics as an inspiration for the philosophical creation of concepts, but we should be very careful with such inspiration to avoid reducing philosophy to the aimless and purposeless profusion of new ideas.

Guattari describes aesthetics as the philosophy of assembly, the integration of parts to constitute a whole (Guattari, 1995, pp. 132–134). Although Freud would appear anomalous in Guattari and Deleuze's list of artistic inspirations of singularization, "What is best in Freud is his literary dimension" (Guattari, 1996, p. 134).[7] In crafting his interpretations of dreams, Freud makes himself a psychoanalytical artist. This vision of Freud illustrates how aesthetic creation can arise in any discipline and context of thought. Wherever bodies and processes interact to constitute larger and more complex bodies and processes, such assembly is art. Such artwork needs no artist directing its creation. Processes collide to assemble themselves into larger networks of processes, an aesthetic creation without director. Relation itself has productive power.

The title of Guattari's last solo book, *Chaosmosis*, has an artistic inspiration, James Joyce's *Finnegans Wake*: "every person place and thing in the *chaosmos* of Alle anyway connected with the gobbly-dumped turkery was moving and changing every part of the time" (Joyce, 2000, p. 118).[8] The subtitle of *Chaosmosis* is *An Ethico-Aesthetic Paradigm*. It is reasonable for us to conclude from this that this book illustrates how aesthetics and ethics are two aspects of the same philosophical domain. The passage from Joyce suggests that this conception of aesthetics is the best framework to understand reality without having to rely on stable categories of thinking. Aesthetics of directionless production understands the universe not as it is at any one moment, but how it becomes from moment to moment, across durations (Guattari, 1995, p. 109). At first glance, you would think this would be an abuse of the term "aesthetic," because we usually reserve terms like "artful" for carefully crafted products. But this conception of aesthetic knowledge has expanded its domain from a restriction to humanist contexts of artistic craft to become a philosophical science of assembly itself. Aesthetic analysis concerns how bodies are assembled. The aesthetics of an assemblage is the history and structure of its assembly, and we can learn a great deal from such an analysis.

However, this broader conception of aesthetics can also let us see the craftsmanship of a human activity that is rarely self-conscious. Environmentalist exemplars are guides for a process that Deleuze and Guattari might call becoming-ecology. Historical examples include Henry David Thoreau, Aldo Leopold, Annie Dillard, and Edward Abbey, among many others (Andersen, 2010, pp. 43–55). The practice of environmental philosophy often encourages the philosophers themselves to live in closer accordance with their principles than many of their readers can manage. John Muir's journals describe a wanderer in the rural California Sierra, a lifestyle that in both Muir's time and today is barely possible for the majority of people (Browne, 2007, pp. 21–23). Nestled in a deep, shining valley of Tvergastein Mountain, Næss's cabin resembles a hermit's retreat, the little house contributing to his public image as a European eco-guru. The cryptic style of his speech during interviews about his environmentalist lifestyle suggests the Bavarian prophet Mühlhiasl reborn as an ecologist (van Boeckel, 1997).

Giving up so much technological convenience, the last century's environmentalist exemplars aimed to inspire others to change their lives along a more ecologically sustainable trajectory. In this way, an environmentalist makes of her own life an intentional kind of artwork. She shapes her lifestyle and personality according to her philosophical principles, an *ethical* process. In crafting her lifestyle, she also carefully crafts her public image, an *aesthetic* process that serves a *political* aim of advancing the environmentalist movement as a role model. Using this book's conception of subjectivity, these biographies can work as exemplars. However, the guidance of exemplars like Næss in Tvergastein can stumble on the singularity of human personalities. This problem underlies Scott Aikin's critique of the deep ecologist's intuition of the intrinsic value of all things, though his analysis stuck to the fact of its impossible universality and did not seek the reasons. He is right in saying that one cannot discover universally necessary moral principles through the intuition arising as one "sees a big rock and has an experience" (Aikin, 2008, p. 27). The experience and its provocation to think is peculiar to that singular experiencer. However, exemplary experiences and lifestyles are more than simply intuition pumps. We cannot all move to a cabin in rural Norway and become mountaineers. Each of our thoughts and actions must constitute our own narrative, laying down a path by walking it (Varela et. al, 1991, pp. 144–145).

An exemplar inspires a creative subjective process, breaking down and reassembling a subject into a new subjectivity. Exemplars provide a guide to singularization, not a model to imitate. Imitation is becoming *like* a rat or *like* a woman. You take your model's characteristics, but not his character, becoming his indistinguishable double with enough skill and attention. Imitating a model trades one totalizing subject structure for another, and does not singularize a personality at all. But imitation produces only doppelgangers, unfortunately precise. A guide is no idol, instead sparking your own lifestyle's transformation into a new subjectivity. Such a movement is within the power of your self-conscious control, but much also is beyond it. Our new habits and desires, on reaching an idiosyncratic critical point in their development, continue spontaneously. A guide is the first to transform, through which she indicates a direction that has never been taken before. With the inspiration of a guide, you can transform yourself without necessarily resembling your guide. You and your guide take the same direction by different paths. Exemplars are the ultimate political activists in that their own lives become artworks to inspire others.

Deleuze and Guattari's account of how exemplars work also rejects the class hierarchy implicit in Næss's conception of the difference between philosophers and ordinary people. Næss understood philosophically and environmentally progressive people as a kind of benevolent police. They keep the masses, whose thinking can be no more complex than simple slogans, in line with environmentalist norms. However, any self-conscious body can break from social norms that have become obsolete, trying to inspire others to transform themselves along similar trajectories. The structure of subjectivity includes the power to singularize itself into an exemplar, a function of the collision of its individual and social processes. Those lucky ones who become noteworthy enough to inspire many people spark a mass movement in creating a new kind of subjectivity, and their own lives are successful experiments (Deleuze and Guattari, 1987, p. 291). Philosophy and art converge on the singularization processes of actual subjectivities in the guiding activities of exemplars. These processes unsettle and overthrow old models of thinking, and have the potential to create a new epoch in thought and society. Even though the change may happen slowly, sometimes over many generations, a population can find little in common with its ancestors from before the social and moral paradigm shift. Whether in thought, society and politics, or ecosystemic assemblages, the activities of their constituents create a new way of life (Stengers, 2010, p. 52).

Conclusion: Intrinsic Value in Ethics and Morality

Two principles about the nature of existence lie at the heart of the conceptual transformation of humanity that this book describes: (1) Thanks to their common origin in the beginning of the universe itself, the processes that generate all bodies in the universe also integrate those bodies into a multiplicity unified by relation; (2) All bodies are singular in the absolute sense of their uniqueness, and the degree and manner of a body's divergence from spatially and temporally nearby processes. Adapting these principles to the political and social goals of environmental philosophy makes belief in them imply a moral imperative to safeguard and encourage diversity. We ensure that we will follow a moral imperative by cultivating our own personalities accordingly. I am more likely to act in a particular way if such action flows from my own nature. I don't have to force myself to act, if I desire that action. Just as it is easier to be kind when the happiness of others pleases us, it is easier to work for natural diversity when a diverse world brings us joy. Not every environmentalist exemplar will successfully inspire such a widespread transformation of humanity to constitute revolutionary social change. But the uncertainty of success is no reason never to try.

The political program of environmental philosophy requires a new kind of human subject who can care for an object because it is different. From the fact of their singularity in the absolute, each HIV virus, each genocidal dictator, each toxin-spewing chemical plant is valuable, despite their destructiveness. Life is impossible without destruction. All change is destruction because a singular state of affairs passes away, never to exist precisely the same way again. Yet everything changes; generation is both creative and destructive. Process creates the new, its condition being the destruction of what once was. Each act of change permanently forecloses on innumerable other events that might otherwise have been. This is the existentially tragic dimension of process philosophy. Whatever comes to be is singular, so is valuable absolutely. Understanding singularities' absolute value brings joy to even the most mundane elements of existence. However, what passes away is also singular and valuable absolutely, so mourning and sadness accompany joy in singularity.

Humanity's survival is caught in this double bind, because even an ecologically sustainable lifestyle will destroy absolutely valuable assemblages. Each of us, and humanity as a whole, are also absolutely valuable processes. So ending your life will not prevent the destruction of the absolutely valuable. The practical dimension of difference, relative benefits and harms, constitutes moral relevance because we can act on it. Ecology's focus on the interdependence of processes in their generation and transformation provides our framework for practical action: identifying which processes will empower or harm us. Investigating what relations empower and harm us and other processes builds our practical understanding of how singularity works in the world. Our investigations are inevitably partial to some degree, always oriented to some specific problem, disciplinary domain, or arena of practical action. Næss's science of virtues that distinguished egocentrism from the awareness of interdependence guides our partial perspectives to the best outcome. Contemporary enormous industry is the most prominent and destructive product of our egocentric and shortsighted thinking on a civilizational scale.

A key contribution of environmental moral philosophy to contemporary thinking is its detailed analysis of enormous industry. Valuing absolute singularity gives us a bulwark against egotistical thinking, guiding our practical actions to keep sight of the long-term and long-distance consequences, as well as our actions' significance. Such an attitude keeps us mindful of the fatal harms that can emerge from immediate technological benefits. Take nothing for granted because no one body or process is exactly like any other. From an existential perspective, life is simultaneously and paradoxically joyful and mournful because process both creates and destroys, and in its absolute value, what is destroyed is just as worthy of existence as what comes to be. This paradox of joy and sadness does not disappear because attitudes of absolute valuation have practical benefits. The tension of the paradox's two sides is ultimately productive, creating our path to appreciate the significance of change.

This book's vision a new humanity is a humanity that is aware of its limits. The structure of our perceptual apparatus and linguistic systems shapes and restricts our capacities for sensation, knowledge, and communication. The limits of our knowledge imply ethical and moral principles about human self-conception and action. The processes that generate the universe at every moment are immensely complex. So any simple statement about some aspect of this vast and complex universe inevitably distorts the totality of facts so we can understand them easily. This is why, when Merleau-Ponty turned to his favourite painter Paul Cézanne, for examples of how to understand reality's complexity, he wrote that any description of his artwork offers us several Cézannes (Merleau-Ponty, 1962, p. 174). General propositions can never be fully adequate to the actual phenomenon we use them to describe (Al-Saji, 2008, p. 53). The existential dilemma of Sartre's *Nausea* asks why we should strive for knowledge at all, given the inadequacy of human language, the vehicle of our scientific knowledge, to give an exhaustive account of reality. Accepting the inadequacy of our knowledge to comprehend all that is suggests an attitude of humility and a moral posture of caution. No matter how comprehensive and precise our understanding, some detail may still escape our knowledge. That elusive fact may make the difference between a vital process and a catastrophe. We can imagine a totally complete system of all possible knowledge, our most sorrowful history and literature is filled with the tragic folly of those who believed they had achieved it.

Building an attitude of humility in our technological pursuits does not require an intuition of nature's sublimity in experience, whether of immensity in hiking the Andes or complexity in calculating the possible variations in protein folding. We need only understand the epistemic principle of the limits of language and thought to render the total complexity of existence. Our knowledge can improve or degrade, but there is intolerable hubris in believing its perfection. Mindfulness of our limitations and finitude produces humility without having to postulate any sense of mystery to nature or its inherent alienation from us. The intuition of nature's value in experience too often takes on a mystical or religious character to humble us with its overwhelming mystery (Kohák, 1984, p. 117). Our better alternative is simply understanding the mystery of nature to lie in its power to surprise us (Meillassoux, 2008, pp. 84–86). The inevitably surprising character of the universe also follows from the autopoietic aspects of the human organism. The boundary of an autopoietic body separates its inside from its environment, and the environment is inevitably more complex. So each of us are always playing catch-up to the wider universe, as internal information always has a different character than the external phenomena it describes (Wolfe, 2009, p. 220). A body's inside is a network of relations that form multifaceted catacombs from its biochemical structures and flowing complexities of thought and memory in a human personality (Deleuze, 1993, p. 1993). We will always find our world more obscure than we want, and this fact must inform our caution and humility.

These moral and ethical considerations overcome the self-destructive stumbling of technologically powerful organisms acting from a shortsighted egoistic self-interest. This is the same goal that Næss wanted his dogma centred on the concept of self-as-place to maintain. It can easily reduce to the precautionary principle, and Næss considers the apparent obviousness of this idea that we should be careful when we control great power a philosophical virtue. On his thinking, the masses can only take on a principle if it has a clear, intuitive basis. But just because an idea has become a platitude does not make it valid or true. The precautionary principle is not a philosophical foundation, but a means to an end. You have already decided that humanity is worth saving, and adopting the precautionary principle for your everyday thinking can help you achieve this. The actual first principle of ecological philosophy emerges from how we work through the dilemma of whether humanity should continue existing at all. My answer is that humanity is a remarkable singularity in nature, and we can also promote and encourage the singularization. Ecocentric moralities, which understand singularization as an imperative to increase and safeguard diversity, are best suited to this task.

The precautionary principle is the only moral imperative that an ecocentric philosophy can coherently hold regarding human technology. Although it was in vogue in environmentalist moral philosophy for many years, the principle of nature's intrinsic value can only function coherently regarding problems of ethics and ontology. It offers a conception of humanity as a singular, remarkable creation of nature. Such a vision of the human overcomes any misanthropy that may arise from encountering the punishing sorrow of environmental destruction. Yet our new human remains in paradoxical tragedy, simultaneous joy and mourning at the world's and her own existence. Creation depends on loss, but humanity has become the agent of so much loss that it dwarfs the capacity of the most dedicated memorial to the extinct.

Existence itself is a memory, if of a different kind that humanity's record keeping. All processes literally continue in their proliferating affects, their having changed all the interactions through which they related to everything else in the universe. All that has ever been remains part of what is because their long-ago activity helped constitute the current state of affairs. Articulating this ontological point in an ethical context creates a metaphysical principle that univocally affirms existence. As such the principle of nature's intrinsic value, while not properly moral itself, is the foundation of an ecocentric morality. This book has described a subjectivity uniquely suited to an ecocentric morality. To renew the world after the ecological damage over our current epoch as a civilization, the new humanity must consider itself integrated with its environment in the most essential facets of its generation and existence.

The new human will conceive of herself as a place, when our necessary conceptions of a subject itself will not be independence, but interdependence. We can still slip too easily into the habit of thinking that the isolation of the self is obvious (Guattari, 2008, p. 28). Our habits only hold us back from the progress on which humanity's survival depends. For the new human, it would be an obvious truth that all bodies are assemblages of other bodies and fields of force, and their constitutive relations are processes generated in the continuous chain of processes that is the universe's history. The world is complex and plural, but also unified. Adequately understanding the world cannot focus on one aspect of existence and ignore others (Guattari, 1996, p. 121). Changing a socioeconomic order, for example, is not an independent task from ecological shifts and new models of individual psychologies. Satisfying the political demands of a previously disenfranchised group is not a genuine revolution, only the concession of one force to another. The system through which those social forces relate is conserved. Real revolution changes the entire framework in which individuals and groups articulate and understand their demands and their identities. To create a new model of human life is inherently a political revolution.

I refer to a civilizational moment of posthumanism as a revolution, but genuine revolution is a much quieter affair than our violence associations with the word suggests. In an interview from 1985, Guattari explains this point through the example of a political problem, how to achieve economic security for workers. Genuine economic security for workers does not depend on wages alone; higher wages instead become an excuse for corporate managers to move factories to regions where workers are paid less. The better solution for workers, says Guattari, is to make your demands in the language of guarantees. Specific wage levels themselves are less important than the security of the social place workers occupy, the guarantee that their society will continue to support them, and value sustaining a community's economy over the quick hits of immediately higher profits that come from global labor market mobility (Guattari, 1996, pp. 127–130). Workers must transform how they seem themselves, which is a matter of how they understand who and what they are. Genuine political change is the proliferation of different kinds and methods of thinking among many people. All revolution, whether socialist, ecological, or whatever, can only succeed where it is first a revolution in thought, in understanding.

Ecological philosophy's revolution in thought is unlike any previous political revolution in human history, insofar as its new framework of thinking opposes humanism. An ecological human would stand against the privilege of humanity as a special element of existence, more special than anything else that exists. My own vision of what our posthuman future must be if humanity is to have a future is similar to many of the philosophies that I critiqued throughout this book. Consider how Peter Singer's concept of speciesism attacks the humanist moral traditions that bestow privilege on humanity to inflict suffering on nonhumans. Ever since people began identifying themselves as environmentalists, and even before them, people have stood against humanity's hubris. But it is not enough to attack hubris; we must replace it.

Wholesale replacement must leave no element of humanity's self-conception untouched. A new vision of humanity must overturn and rebuild our ontological, epistemic, moral, and ethical domains of thinking, action, and daily life. Guattari calls such a comprehensive transformation an "ecological industry" that remakes and reorganizes civilization from a humble perspective (Guattari, 1995, pp. 123–124). Philosophy does not create specific rules for the political and industrial program of such a civilization and deliver it to the people like a set of orders. Philosophy is an educational discipline, guiding humanity toward posthumanity, and so on again as this new humanity becomes obsolete or maladaptive. Ecological philosophy is a guide to develop a sustainable technological civilization. An ecological human is a humble self that understands her interdependence with the world through her self-conception as her place. Such a person's intelligence works to encourage sustainability and explore the fascinating complexity of existence.

NOTES

1 Nature's Intrinsic Value: A Forgotten Philosophy of the Environment

1. Think of this concept of intrinsic value as a moral corollary of Leibniz's Law, when we understand Leibniz's Law of the identity of indiscernibles as an ontological principle.
2. Writers such as Bryan Norton, Ben Minteer, and Andrew Light, who identify as environmental pragmatists, have no problem with valuation originating in human thought, but the debate between the two sides has come down to an impasse over the last decades.

2 Looming Ecofascisms in the Value of Nature

1. Spinoza's goal in this passage is to describe an error of reasoning. For him, it is a terrible mistake to conceive of humanity as a separate category from nature, as an exemption from nature's laws and systems, as being somehow above or beyond nature.
2. As Kenneth Mars said in *The Producers* (1968), Hitler was a wonderful dancer. This is no reason to denounce dance.
3. The social, political, and philosophical transformations that the First World War ushered into Western civilization has far different ramifications than what applies directly to environmental politics and ecological philosophy. Writers like Jünger and the Italian Filippo Marinetti created a terrifying vision of a mechanized utopia that was as much a profanation of humanity as it was of Earth as a whole. But examining this will be another project.

3 Two Paradoxes of Practical Philosophy

1. Callicott's hostile reading of pragmatist philosophers such as William James and John Dewey might turn readers away from productive ideas found in a more charitable interpretation of these thinkers. See, for example, Andrew Light's *Environmental Pragmatism*, and Adam Riggio's "John Dewey as a Philosopher of Contingency and the Value of this Idea for Environmental Philosophy," referenced in this volume.

2. "Not Man Apart!" is the slogan of the Sierra Club, and "The Future Is in Our Hands!" is the name of a Norwegian environmentalist political movement.
3. As a young man, David Foster Wallace was the only educated person in his addiction recovery group, and was shocked and impressed to find corny slogans like "One day at a time," sparking transformations in the lives of his fellow addicts. He began to see himself as burdened by his education in contemporary philosophy and work in postmodern literature. So used to nuanced analysis, he found it more difficult than his less-educated fellows to learn the mental habits of recovery.
4. In this case, public relations messaging is another area in which Næss was oddly prescient.
5. The "T" stands for Tvergastein Mountain, the location of the hut where Næss composed many of his later philosophical and political writings. He intends the initial to indicate the personal, idiosyncratic nature of an ecosophy, because in his vision of an ecologically mindful society, every individual would have an ecosophy of their own, all including norms focused on sustainability.
6. Kohák points out that the slower deaths by starvation of Grand Canyon horses caused no public outcry, because the media only covered the morally simple tale of human hunters shooting down majestic animals. This image is an example of the folly of thinking according to the absolute dichotomy of human and nature, and Kohák's story provides an example of the mass media's role in producing the theatricality required to encourage political mobilization.
7. Yes, this does contradict his endorsement of philosophers as an ecocentric thought police. It is impossible to treat Næss's environmental philosophy as an internally consistent system precisely because many principles, such as the conception of philosopher as environmentalist morality police and the concept of the ecocentric exemplar, are incompatible. I don't approach Næss as offering a coherent system to be adopted as a whole, but as a collection of principles that I can evaluate on their own and adapt to my own thinking or discard as I wish. You can do that to me.
8. I consider the term "total view" problematic because of the connotations of the word "total." I use the terms "totalization" and "totality" throughout this work to refer to a conception of comprehensiveness, which implies that one's knowledge of all existing and possible states is complete, and that nothing can surprise a totalizing system of understanding. Totalization is a conceptual version of how Leopold described the monoculture farm: a single concept that is used to eradicate all possible forms of thinking but those compatible with the concept itself. Just as monoculture farming weakens the physical health of the land, totalizing thought weakens the mental health of a person. I conceive of a comprehensive worldview as a framework of thinking that one can apply to any situation, but which makes no demands that a situation in the world should be understood only in terms of

that framework. One should always allow, as a matter of prudence at least, different perspectives to understand the world.

4 The Essence of an Ecological Philosophy

1. The biochemical processes constitutive of a human body can be objects of self-conscious decision making, as in decisions affecting one's health, like whether to smoke, or begin taking some pharmaceutical drug. But this is an example of how a person's power to think can affect processes that began long before that power was developed.

5 The Conditions of Selfhood

1. Varela began working with Maturana as a graduate student researcher in Santiago in 1966, the beginning of a close partnership that would last until 1973, when both men had to flee Chile after the Pinochet coup. Elsewhere in this essay, Varela describes himself as a "militant supporter of President Allende's government." Varela and Maturana would continue to collaborate on research after settling outside Chile, coauthoring articles and books until Varela's death in 2001. Varela's own training included not only biological science, but also philosophy. He cites philosophers Georges Canguilhem, Gaston Bachelard, Alexandre Koyré, and Thomas Kuhn as seminal writers in shaping his conception of what science could be. The phenomenological tradition of Husserl, Heidegger, and Merleau-Ponty was also important for the long-term development of his biological ideas about autopoiesis, and what this meant for subjectivity.
2. Andy Clark's theory of embodied mind can be considered as a modification of the enactive theory. His theory of mind, where mental operations include perceptual processes that take place outside the human body, is described in Andy Clark, *Supersizing the Mind: Embodiment, Action, and Cognitive Extension* (Oxford: Oxford University Press, 2008). In my view, a decisive critique of Clark's approach can be found in Ezequiel DiPaolo, "Extended Life," *Topoi: An International Review of Philosophy* 28 (2009): 9–21. DiPaolo shows that, even though Clark pitches his extended mind hypothesis as a radical departure from received concepts of mind, Clark presupposes many of the mainstream presuppositions about what thinking is that he says should be discarded.
3. Fodor is a nakedly partisan advocate for the compositional theory, but he makes the distinction in very clear language, despite the presumptuous tone he takes toward those with whom he disagrees.
4. I find Næss's late-period dismissal of argumentative rigor ironic, given his early period affiliation with the Vienna Circle.
5. Remember that these incompatibilities of interest do not ever go away at the individual scale in actual practice, no matter how profound your

self-realization becomes. Self-realization is a matter of emphasis: whether you consider ego or interconnectivity more important.

6. Legal continuity is also of key importance, but I interpret this as a moral and political continuity rather than ontological. DeLanda leaves legal matters aside in this example, and so will I.
7. Næss is keen to clarify that he does not consider this what Spinoza himself meant; it is his own uptake of a Spinozist conception of a person's nature. We need such a clarification to take his conceptual inheritance from Spinoza seriously. Næss's inspiration would have little credibility if he pitched it as what Spinoza *himself* thought, because he would have to interpret the straightforward text of Book I, Definition 3 as being completely different from what it actually says. Næss is not interested in getting Spinoza right, in the strict sense, but instead using Spinoza for his own contemporary philosophical inquiries. This is also how I use Næss, and just about every other major philosophical source of this project.
8. Peter Singer's notion that the desire to avoid suffering constitutes an interest that all animals share also challenges this humanist consensus, but he and other animal rights and liberation philosophers do not approach a deep ecologist's radicalism.
9. Understanding a body's activities as the articulation of interests is not a device exclusive to Næss, but occurs in Plumwood's work as well. I concentrate on Næss because his attempt to apply ecological philosophy directly to political activism falls flat in a way that lets us learn a great deal from the wreckage. Plumwood's writing sticks to the disciplinary domain of philosophical thought.

6 Discovering Active Nature in the Subject

1. In his working notes for the remainder of the book, never finished thanks to his sudden death in 1961, Merleau-Ponty describes phenomenology as trying "to disclose a non-explicated horizon" of experience, understanding of which requires "taking possession of the world of silence." He refers to the paradoxical problem of trying to write in a book an element that is not even directly perceivable in experience, let alone articulable in language.
2. You can describe an organism as a sensible sentient because of the rudimentary perceptual capacities of all autopoietic systems. Directional movement requires some rudimentary proprioception: an organism must have some sense of its physical body to move itself and interact with its surroundings. This sensible sentience reaches an extremely high level of intensity and complexity in human self-consciousness. Of course, it would be quite hubristic of us to think that human self-consciousness was the highest level possible of intensity and complexity.
3. My translation; Lingis renders "Some for the others and not only each for the other," which I think misses Merleau-Ponty's developing a concept of perception as a field of multiplicities.

4. The relation of visible to invisible is a matter of contention in scholarship on late Merleau-Ponty, which is why I am clear that my own speculations on the matter are about how I, not he, intend to use the concepts. That Merleau-Ponty never spelled out this matter in *The Visible and the Invisible* is an unfortunate casualty of his premature death. I find most frustrating about studying Merleau-Ponty that the awful timing of his sudden death left his most interesting book a fragment of what it could have been.
5. It is an idle speculation on my part, but I do wonder if a hominid with even greater neuroplasticity than ours would appear had our genus continued to diversify instead of dwindling to a single species, *homo sapiens*.
6. Michael argues, and I agree, that the environmentalist concept of good as following nature does no important theoretical work: in terms of practical action in the world, the moral principle to follow nature inevitably amounts to a principle of minimizing or avoiding harm to generally healthy ecosystem-constitutive processes. An ideal image of nature (Gaia, for example) has no practical purpose when mindful care does the same job.
7. Uniformity and routinization of production results in higher short-term efficiency no matter what the sector. Throughout his book, DeLanda discusses agriculture, market exchanges, and industrial manufacturing as examples.
8. If you conclude that humanity is inherently technological, and that technology is an inherently destructive, exploitive, rapacious, and so evil force, then your most morally good action is to commit suicide. I, meanwhile, will be alive, still trying to solve the problem of survival.
9. Uexküll is a historical precedent for a concept of subjectivity that is not exclusively human.

7 Ecological Selfhood, Ecological Life

1. This book was Næss's most systematic treatment of the self-as-place concept. In the rest of his ecophilosophy corpus, he writes in essays designed for activist public consumption, or contextually already restricted to moral philosophy proper. The former oversimplifies his ideas into motivational exhortations and rallying cries. The latter relegates ontological concepts to the background while spelling out the economic and political applications of the normative principles.
2. It is also possible, given the convergence this chapter's argument implies of late-period Merleau-Ponty with the work of Deleuze and Guattari, to understand the stable subject presupposed as the possibility condition of all intentional action to be the overcoding of the body by faciality in the context of *A Thousand Plateaus*, and the universality of phenomenology's transcendental ego/I in the context of Merleau-Ponty's Husserlian legacy. But this particular task might be more difficult.
3. While contemporary research provides evidence that the neurological cause of schizophrenic behavior is a decrease in the activity of important infrastructure of the brain, what is important for my argument is what Guattari

was able to think of thanks to the worldly behavior of schizophrenics, the disinhibited behavioral expression of this inhibition of neurological activity.

4. In Deleuze and Guattari's vocabulary, the philosophy of departure deterritorializes the thinking of its readers, who reterritorialize thought along this new model by thinking and writing according to its problems instead of the old problems. Although the context varies considerably, this parallels the phenomenon that Rorty called the shift in a community's vocabulary, with all the promise and risk any new vocabulary entails. It also parallels Thomas Kuhn's concept of a paradigm shift in a scientific community.
5. Manuel DeLanda, personal correspondence, June 24, 2011.
6. Rorty divests historical developments of any importance because they depend on contextual circumstances; they "just happened that way." His contingency is deflationary, while Deleuze and Guattari's is empowering. With no force of necessity dictating historical development, everything that constitutes the Earth shares responsibility for the events of history.
7. I understand this as the central critique of psychoanalysis in *Anti-Oedipus*: Freud was primarily a literary interpreter of human personalities, turning the psychological narratives of his patients into insightful images, just as singular as his patients themselves. Psychoanalysis as a discipline, even Freud himself eventually, mistook his singular interpretations of individual cases for the universally necessary structure of human subjectivity that could be applied without modification to all people.
8. My emphasis. Scientists are also inspired by art, as the name of the fundamental particles whose interactions constitute nucleic particles like protons and neutrons, were named quarks, "Three quarks for Muster Mark!" (383).

Bibliography

Aaltola, Elisa. "Personhood and Animals." *Environmental Ethics* 30, no. 2 (Summer 2008): 175–194.

Abram, David. "Between the Body and the Breathing Earth: A Reply to Ted Toadvine." *Environmental Ethics* 27, no. 2 (Summer 2005): 171–190.

Abram, David. "Merleau-Ponty and the Voice of the Earth." *Environmental Ethics* 10, no. 2 (Summer 2008): 101–120.

Abram, David. *The Spell of the Sensuous: Perception and Language in a More-Than-Human World*. New York: Pantheon Books, 1996.

Abrioux, Yves. "Intensive Landscaping." In Herzogenrath 2009, 251–265.

Adams, Carol. J. "The Feminist Traffic in Animals." In *Ecofeminism: Women, Animals, and Nature*. Edited by Greta Gaard, 195–218. Philadelphia: Temple University Press, 1993.

Adams, Emily D. M., Greg G. Goss, and Sally P. Leys. "Freshwater Sponges Have Functional, Sealing Epithelia with High Transepithelial Resistance and Negative Transepithelial Potential." *PLoS ONE* 5, no. 11. November 29, 2010. Accessed December 12, 2010. e15040. doi:10.1371/journal.pone.0015040

Agar, Nicholas. *Life's Intrinsic Value*. New York: Columbia University Press, 2001.

Aikin, Scott F. "The Dogma of Environmental Revelation." *Ethics and the Environment* 13, no. 2 (2008): 23–34.

Ales Bello, Angela. "'Brute Being' and Hyletic Phenomenology: The Philosophical Legacy of Merleau-Ponty's *The Visible and the Invisible*." In Tymieniecka 2009, 55–76.

Alexander, Joshua, Stacy Swain, and Jonathan M. Weinberg. "The Instability of Philosophical Intuitions: Running Hot and Cold on Truetemp." *Philosophy and Phenomenological Research* 76, no. 1 (January 2008): 138–155.

Al-Hassan, Ameera H., and Steven J. M. Dudek. "Teaching Sustainability in Schools of Architecture: Prescribed or Integrated?" Paper presented at the Sixth International Conference on Environmental, Cultural, Economic, and Social Sustainability, Cuenca, Ecuador, January 5, 2010.

Allen, Barry. *Knowledge and Civilization*. Boulder: Westview Press, 2004.

Al-Saji, Alia. "'A Past Which Has Never Been Present:' Bergsonian Dimensions in Merleau-Ponty's Theory of the Prepersonal." *Research in Phenomenology* 38 (2008): 41–71.

Al-Saji, Alia. "The Temporality of Life; Merleau-Ponty, Bergson, and the Immemorial Past." *Southern Journal of Philosophy* 45, no. 2 (2007): 177–206.

Andersen, Nathan. "Exemplars in Environmental Ethics: Taking Seriously the Lives of Thoreau, Leopold, Dillard, and Abbey." *Ethics, Place, and Environment* 13, no. 1 (2010): 43–55.

Anderson, Christopher B., Gene Likens, Ricardo Rozzi, Julio R. Gutiérrez, and Gutiérrez & Juan J. Armesto. "Integrating Science and Society through Long-Term Socio-Ecological Research." *Environmental Ethics* 30, no. 3, 295–312.

Andreou, Chrisoula. "A Shallow Route to Environmentally Friendly Happiness: Why Evidence That We Are Shallow Materialists Need Not Be Bad News for the Environment (alist)." *Ethics, Place, and Environment* 13, no. 1 (2010): 1–10.

Appiah, Kwame Anthony. *Experiments in Ethics*. Cambridge: Harvard University Press, 2008.

Asimov, Isaac. *Foundation and Earth*. New York: Doubleday, 1986.

Atkinson, Michael. "Parkour, Anarcho-Environmentalism, and Poiesis." *Journal of Sport Issues* 33, no. 2 (May 2009): 169–194.

Attfield, Robin. "Social History, Religion, and Technology: An Interdisciplinary Investigation into Lynn White Jr.'s 'Roots.'" *Environmental Ethics* 31, no. 1 (Spring 2009): 31–50.

Attfield, Robin. "The Shape of a Global Ethic." *Philosophy and Social Criticism* 32, no. 1 (2006): 5–19.

Ayob, Gloria. "The Aspect-Perception Passages: A Critical Investigation of Köhler's Isomorphism Principle." *Philosophical Investigations* 32, no. 3 (July 2009): 264–280.

Badiou, Alain. *Wittgenstein's Antiphilosophy*. Translated by Bruno Bosteels. New York: Verso, 2011.

Baerveldt, Cor, and Paul Voestermans. "Culture, Emotion, and the Normative Structure of Reality." *Theory and Psychology* 15, no. 4 (2005): 449–473.

Baichwal, Jennifer. *Edward Burtynsky: Manufactured Landscapes*, DVD. Directed by Jennifer Baichwal. 2006. Toronto: Mercury Films, 2006.

Baldwin, Jeff. "The Culture of Nature through Mississippian Geographies." *Ethics and the Environment* 11, no. 2 (2006): 13–44.

Bankowski, Zenon. "How Does It Feel to Be on Your Own? The Person in the Sight of Autopoiesis." *Ratio Juris* 7, no. 2 (1994): 254–266.

Barbaras, Renaud. "Francisco Varela: A New Idea of Perception and Life." *Phenomenology and the Cognitive Sciences* 1 (2002): 127–132.

Barbaras, Renaud. "Merleau-Ponty and Nature." Translated by Paul Milan. *Research in Phenomenology* 31 (2001): 21–38.

Bechtel, William, and Adele Abrahamsen. "Explaining Human Freedom and Dignity Mechanistically: From Receptive to Active Mechanisms." *Journal of Philosophical Research* 32 (2007): 43–66.

Bekoff, Marc, Jan Nystrom. "The Other Side of Silence: Rachel Carson's Views of Animals." *Zygon: Journal of Religion and Science* 39, no. 4 (December 2004): 861–884.

Benjamin, Walter. "The Work of Art in the Age of Mechanical Reproduction." In *Illuminations*. Edited by Hannah Arendt, 217–251. New York: Schocken Books, 1968.

Bennett, Jane. *Vibrant Matter: A Political Ecology of Things*. Durham: Duke University Press, 2010.

Benovsky, Jiri. "The Bundle Theory and the Substratum Theory: Deadly Enemies or Twin Brothers?" *Philosophical Studies* 141, no. 2 (November 2008): 175–190.

Berressem, Hanjo. "Structural Couplings: Radical Constructivism and Deleuzian Ecologics." In Herzogenrath 2009, 57–101.

Bigwood, Carol. "Renaturalizaing the Body (with the Help of Merleau-Ponty)." *Hypatia* 6, no. 3 (Autumn 1991): 54–73.

Bogue, Ronald. "A Thousand Ecologies." In Herzogenrath 2009, 42–56.

Bookchin, Murray. "Toward a Philosophy of Nature: The Bases for an Ecological Ethics." In Tobias 1984, 213–239.

Bourdieu, Pierre. *Homo Academicus*. Translated by Peter Collier. Stanford: Stanford University Press, 1988.

Bratman, Michael E. "Shared Intention." *Ethics* 104, no. 1 (October 1993): 97–113.

Bratton, Susan Power. "Luc Ferry's Critique of Deep Ecology, Nazi Nature Protection Laws, and Environmental Anti-Semitism." *Ethics and the Environment*, 4, no. 1 (1999): 3–22.

Briggs, Roman. "The Greening of Heart and Mind: A Love Story." *Environmental Ethics* 31, no. 2 (Summer 2009): 155–168.

Brigham, Linda. "Communication Versus Communion in Psychic Systems: Maturana, Luhmann, and Cognitive Neurology." In Clarke and Hansen 2009, 205–219.

Brighenti, Andrea Mubi. "On Territorology: Towards a General Science of Territory." *Theory, Culture, and Society* 27, no. 1 (2010): 52–72.

Brodsky, Joyce. "A Paradigm Case for Merleau-Ponty: The Ambiguity of Perception and the Paintings of Paul Cézanne." *Artibus et Historiae* 2, no. 4 (1981): 125–134.

Brown, Steven D. "Between the Planes: Deleuze and Social Science." In Jensen and Rödje 2010, 101–120.

Brown, Steven D., and Ian Tucker. "Eff the Ineffable: Affect, Somatic Management, and Mental Health Service Users." In Gregg and Seigworth 2010, 229–249.

Browne, Neil W. *The World in Which We Occur: John Dewey, Pragmatist Ecology, and American Ecological Writing in the Twentieth Century*. Tuscaloosa: University of Alabama Press, 2007.

Buchanan, Brett. *Onto-Ethologies: The Animal Environments of Uexküll, Heidegger, Merleau-Ponty, and Deleuze*. Albany: State University of New York Press, 2008.

Bunge, Mario. *Emergence and Convergence: Qualitative Novelty and the Unity of Knowledge*. Toronto: University of Toronto Press, 2003.

Bunge, Mario. *Social Science under Debate: A Philosophical Perspective*. Toronto: University of Toronto Press, 1998.

Burkitt, Ian. "Psychology in the Field of Being: Merleau-Ponty, Ontology, and Social Constructionism." *Theory and Psychology* 13, no. 3 (2003): 319–338.

Burns, Kelly A. "Warren's Ecofeminist Ethics and Merleau-Ponty's Body-Subject." *Ethics and the Environment* 13, no. 2 (2008): 101–118.

Butler, Clark. "On the Impossibility of Metaphysics without Ontology." *Metaphilosophy* 7, no. 2 (April 1976): 116–132.

Callicott, J. Baird, and William Grove-Fanning. "Should Endangered Species Have Standing? Toward Legal Rights for Listed Species." *Social Philosophy and Policy* 26, no. 2 (Summer 2009): 317–352.

Callicott, J. Baird, William Grove-Fanning, Jennifer Rowland, Daniel Baskind, Robert Heath French, and Kerry Walker. "Was Aldo Leopold a Pragmatist? Rescuing Leopold from the Imagination of Bryan Norton." *Environmental Values* 18, no. 4 (November 2009): 453–486.

Callicott, J. Baird. "Multicultural Environmental Ethics." *Daedalus* 130, no. 4 (Fall 2001): 77–97.

Callicott, J. Baird. "What 'Wilderness' in Frontier Ecosystems?" *Environmental Ethics* 30, no. 3 (Fall 2008): 235–249.

Callicott, J. Baird. *In Defence of the Land Ethic: Essays in Environmental Philosophy.* Albany: State University of New York Press, 1989.

Camus, Albert. "The Myth of Sisyphus." In *The Plague, the Fall, Exile and the Kingdom, and Selected Essays.* Translated by Justin O'Brien. New York: Alfred A. Knopf Everyman's Library, 2004.

Cannatella, Howard. "Place and Being." *Educational Philosophy and Theory* 39, no. 6 (2007): 622–633.

Carman, Taylor. "Merleau-Ponty and the Mystery of Perception." *Philosophy Compass* 4, no. 4 (2009): 630–638.

Carolan, Michael S. "I Do Therefore There Is: Enlivening Socio-Environmental Theory." *Environmental Politics* 18, no. 1 (February 2009): 1–17.

Carson, Rachel. *Silent Spring.* Boston: Houghton Mifflin, 2002.

Casada da Rocha, Antonio. "The Value of Health in the Writings of H. D. Thoreau." *Environmental Values* 18, no. 2 (May 2009): 201–215.

Catton Jr., William R. "On the Dire Destiny of Human Lemmings." In Tobias 1984, 74–89.

Cheney, Jim, and Anthony Weston. "Environmental Ethics as Environmental Etiquette." *Environmental Ethics* 21, no. 2 (Summer 1999): 115–134.

Cheney, Jim. "Truth, Knowledge, and the Wild World." *Ethics and the Environment* 10, no. 2 (2005), 101–135.

Cheney, Jim. "Universal Consideration: An Epistemological Map of the Terrain." *Environmental Ethics* 20, no. 3 (Fall 1998): 265–277.

Chessa, Frank. "Endangered Species and the Right to Die." *Environmental Ethics* 27, no. 1 (Spring 2005): 23–42.

Chew, Sing C. *World Ecological Degradation: Accumulation, Urbanization, and Deforestation, 3000 B.C.–A.D. 2000.* Walnut Creek: Altamira Press, 2001.

Chipalkatty, Uday. "Signing with Rain for Singing in Rain: Rain Water Harvesting in Urban Areas." Paper presented at the Sixth International Conference on Environmental, Cultural, Economic, and Social Sustainability, Cuenca, Ecuador, January 5, 2010.

Christ, Carol P., and Kathryn Rountree. "Humanity in the Web of Life." *Environmental Ethics* 28, no. 2 (Summer 2006): 185–200.

Churchland, Patricia. *Brain-Wise: Studies in Neurophilosophy*. Boston: MIT Press, 2002.

Clark, Andy. *Supersizing the Mind: Embodiment, Action, and Cognitive Extension*. Oxford: Oxford University Press, 2008.

Clarke, Bruce, and Mark B. N. Hansen. "Neocybernetic Emergence." In Clarke and Hansen 2009, 1–25.

Clarke, Bruce, and Mark B. N. Hansen, eds. *Emergence and Embodiment: New Essays on Second-Order Systems Theory*. Durham: Duke University Press, 2009.

Clarke, Bruce. "Heinz von Foerster's Demons: The Emergence of Second Order Systems Theory." In Clarke and Hansen 2009, 34–61.

Cockell, Charles S. "The Value of Microorganisms." *Environmental Ethics* 27, no. 4 (Winter 2005): 375–390.

Coleman, Frank M. "Classical Liberalism and American Landscape Representation: The Imperial Self in Nature." *Ethics, Place, and Environment* 13, no. 1 (2010): 75–96.

Conley, Verena Andermatt. "Artists or 'Little Soldiers?' Félix Guattari's Ecological Paradigms." In Herzogenrath 2009, 116–128.

Daly, Herman E. "Economics and Sustainability: In Defense of a Steady-State Economy." In Tobias 1984, 90–100.

Dant, Tim. "The Driver-Car." *Theory, Culture, Society* 21, nos. 4/5 (2004): 61–79.

Davidson, Joyce. "A Phenomenology of Fear: Merleau-Ponty and Agoraphobic Life-Worlds." *Sociology of Heath and Illness* 22, no. 5 (2000): 640–660.

Davion, Victoria. "Itch Scratching, Patio Building, and Pesky Flies: Biocentric Individualism Revisited." *Environmental Ethics* 28, no. 2 (Summer 2006): 115–128.

Davis, Mark A., et al. "Don't Judge Species on Their Origins." *Nature* 474 (June 9, 2011): 153–154.

Dawkins, Richard. *The Selfish Gene*. Oxford: Oxford University Press, 1989.

DeLanda, Manuel. "Ecology and Realist Ontology." In Herzogenrath 2009, 23–41.

DeLanda, Manuel. *A New Philosophy of Society: Assemblage Theory and Social Complexity*. New York: Continuum, 2006.

DeLanda, Manuel. *Intensive Science and Virtual Philosophy*. New York: Continuum, 2002.

DeLanda, Manuel. *One Thousand Years of Non-Linear History*. New York: Zone Books, 2000.

DeLanda, Manuel. *Philosophy and Simulation: The Emergence of Synthetic Reason*. New York: Continuum, 2011.

Deleuze, Gilles, and Félix Guattari. *A Thousand Plateaus*. Translated by Brian Massumi. Minneapolis: Minnesota University Press, 1987.

Deleuze, Gilles, and Félix Guattari. *Anti-Oedipus*. Translated by Robert Hurley, Helen R. Lane, and Mark Seem. Minneapolis: Minnesota University Press, 1983.

Deleuze, Gilles, and Félix Guattari. *Kafka: Toward a Minor Literature*. Translated by Dana Polan. Minneapolis: Minnesota University Press, 1986.

Deleuze, Gilles, and Félix Guattari. *What Is Philosophy?* Translated by Hugh Tomlinson and Graham Burchell. New York: Columbia University Press, 1994.

Deleuze, Gilles. "Bartleby; or, the Formula." In *Essays Critical and Clinical*, 68–90. Translated by Daniel W. Smith, and Michael A. Creco. Minneapolis: University of Minnesota Press, 1997.

Deleuze, Gilles. *Cinema 2: The Time-Image*. Translated by Hugh Tomlinson, and Robert Galeta. Minneapolis: University of Minnesota Press, 1989.

Deleuze, Gilles. *Difference and Repetition*. Translated by Paul Patton. New York: Columbia University Press, 1994.

Deleuze, Gilles. *Expressionism in Philosophy: Spinoza*. Translated by Martin Joughin. New York: Zone Books, 1990.

Deleuze, Gilles. *Nietzsche and Philosophy*. Translated by Hugh Tomlinson. New York: Columbia University Press, 2006.

Deleuze, Gilles. *Proust and Signs*. Translated by Richard Howard. Minneapolis: University of Minnesota Press, 2000.

Deleuze, Gilles. *Pure Immanence: Essays on a Life*. Translated by Anne Boyman. New York: Zone Books, 2001.

Deleuze, Gilles. *Spinoza: Practical Philosophy*. Translated by Robert Hurley. San Francisco: City Lights Books, 1988.

Deleuze, Gilles. *The Fold: Leibniz and the Baroque*. Translated by Tom Conley. Minneapolis: University of Minnesota Press, 1993.

Depraz, Natalie. "Phenomenology of Surprise: Levinas and Merleau-Ponty in the Light of Hans Jonas." In Nenon and Blosser 2010, 223–233.

Derbyshire, Alan. "Sustainable Urban Habitats: The Contributory Components of Viable Eco-Diverse Cityscapes." Paper presented at the Sixth International Conference on Environmental, Cultural, Economic, and Social Sustainability, Cuenca, Ecuador, January 7, 2010.

Devall, Bill, and George Sessions. *Deep Ecology: Living as If Nature Mattered*. Salt Lake City: Gibbs M. Smith, 1985.

Diamond, Jared. *Collapse: How Societies Choose to Fail or Succeed*. New York: Penguin Books, 2006.

Diehm, Christian. "Arne Næss and the Task of Gestalt Ontology." *Environmental Ethics* 28, no. 1 (Spring 2006): 21–35.

Diehm, Christian. "Arne Næss, Val Plumwood, and Deep Ecological Subjectivity: A Contribution to the 'Deep Ecology-Ecofeminism Debate.'" *Ethics and the Environment* 7, no. 1 (2002): 24–38.

Diehm, Christian. "Identification: What It Is and Why It Matters." *Ethics and the Environment* 12, no. 2 (2007): 1–22.

Diehm, Christian. "Minding Nature: Val Plumwood's Critique of Moral Extensionism." *Environmental Ethics* 32, no. 1 (Spring 2010): 3–16.

DiPaolo, Ezequiel. "Extended Life." *Topoi: An International Review of Philosophy* 28 (2009): 9–21.

Dosse, François. *Gilles Deleuze and Félix Guattari: Intersecting Lives*. Translated by Deborah Glassmann. New York: Columbia University Press, 2010.

Dostoyevsky, Fyodor. *The Brothers Karamazov*. Translated by Richard Pevear, and Larissa Volokhonsky. New York: Farrar Straus and Giroux, 1990.

Drenthen, Martin. "Fatal Attraction: Wildness in Contemporary Film." *Environmental Ethics* 31, no. 3 (Fall 2009): 297–315.

Duportail, Guy Félix. "Le Moment Topologique de la Phénoménologie Française." *Archives de Philosophie* 73, no. 1 (2010): 47–65.

Dvorak, Laura. "Reclaiming Everyone's Soil-Opportunity to Understand Relational Cycles of Ecology: An Examination of Humanure though the Lens of Global Environmental Crisis and Humanity's Spiritual Catharsis." Paper presented at the Sixth International Conference on Environmental, Cultural, Economic, and Social Sustainability, Cuenca, Ecuador, January 6, 2010.

Edwards, Craig. "Beyond Mental Competence." *Journal of Applied Philosophy* 27, no. 3 (August 2010): 273–289.

Elden, Stuart. "Contributions to Geography? The Spaces of Heidegger's *Beiträge*." *Environment and Planning* 23 (2005): 811–827.

Elliot, Robert. "Instrumental Value in Nature as a Basis for the Intrinsic Value of Nature as a Whole." *Environmental Ethics* 27, no. 1 (Spring 2005): 43–56.

Ereshefsky, Marc. "Psychological Categories as Homologies: Lessons from Ethology." *Biology and Philosophy* 22, no. 5 (2007): 659–674.

Evernden, Neil. *The Natural Alien: Humankind and Environment*. Toronto: University of Toronto Press, 1985.

Ferry, Luc. *The New Ecological Order*. Translated by Carol Volk. Chicago: University of Chicago Press, 1995.

Ferry, Luc; Alain Renaut. *Heidegger and Modernity*. Translated by Franklin Philip. Chicago: University of Chicago Press, 1990.

Fodor, Jerry A. "Having Concepts: A Brief Refutation of the Twentieth Century." *Mind and Language* 19, no. 1 (February 2004): 29–47.

Fodor, Jerry A. *Language of Thought 2: The Language of Thought Revisited*. Oxford: Oxford University Press, 2008.

Ford, Russell. "Deleuze's Dick." *Philosophy and Rhetoric* 38, no. 1 (2005): 41–71.

Fraser, Mariam. "Facts, Ethics, and Event." In Jensen and Rödje 2010, 57–82.

Frodeman, Robert. "Redefining Ecological Ethics: Science, Policy, and Philosophy at Cape Horn." *Science and Engineering Ethics* 14 (2008): 597–610.

Garfinkel, Alan. *Forms of Explanation: Rethinking the Questions in Social Theory*. New Haven: Yale University Press, 1981.

Genosko, Gary. "Subjectivity and Art in Guattari's *The Three Ecologies*." In Herzogenrath 2009, 102–115.

Genosko, Gary. "The Life and Work of Félix Guattari." In Guattari 2008, 46–78.

Geraud, Tatiana, Jes S. Pedersen, and Laurent Keller. "Evolution of Supercolonies: The Argentine Ants of Southern Europe." *Proceedings of the National Academy of Sciences of the United States of America* 99, no. 9 (April 30, 2002): 6075–6079.

Gibson, Glenn R. "Fibre and Effects and Probiotics: The Prebiotic Concept." *Clinical Nutrition Supplements* 1, no. 2 (2004): 25–31.

Gleick, James. *Chaos: Making a New Science*. New York: Penguin Books, 1987.

Goffey, Andrew. "Naturalizing Phenomenology: Cognitive Science and the Bestowal of Sense." *Radical Philosophy* 114 (July/August 2002): 20–28.

Golding, Mark. "Obligations to Future Generations." *Monist* 56, (1972): 85–99.

Graffin, Greg, and Steve Olson. *Anarchy Evolution: Faith, Science, and Bad Religion in a World without God.* New York: HarperCollins, 2010.

Grapard, Alan G. "Nature and Culture in Japan." In Tobias 1984, 240–255.

Gregg, Melissa, and Gregory J. Seigworth, eds. *The Affect Theory Reader.* Durham: Duke University Press, 2010.

Gross, Matthias. "The Public Proceduralization of Contingency: Bruno Latour and the Formation of Collective Experiments." *Social Epistemology* 24, no. 1 (2010): 63–74.

Grossberg, Lawrence. "Affect's Future: Rediscovering the Virtual in the Actual." In Gregg and Seigworth 2010, 309–338.

Guarner, Francisco, and Juan R. Malagelada. "Gut Flora in Health and Disease." *The Lancet* 361, no. 9356 (February 2003): 512–519.

Guattari, Félix. *Chaosmosis: An Ethico-Aesthetic Paradigm.* Translated by Paul Bains and Julian Pefanis. Indianapolis: Indiana University Press, 1995.

Guattari, Félix. *The Guattari Reader.* Edited by Gary Genosko. Oxford: Blackwell Publishers, 1996.

Guattari, Félix. *The Three Ecologies.* Translated by Ian Pindar and Paul Sutton. New York: Continuum, 2008.

Hacking, Ian. *The Social Construction of What?* Cambridge: Harvard University Press, 1999.

Hall, Matthew. "Plant Autonomy and Human-Plant Ethics." *Environmental Ethics* 31, no. 2 (Summer 2009): 169–181.

Halsey, Mark. "Deleuze and *Deliverance*: Body, Wildness, Ethics." In Herzogenrath 2009, 233–250.

Handwerk, Brian. "Photos: Giant Ocean-Trash Vortex Documented—a First." *National Geographic News*, September 4, 2009. Accessed September 5, 2010. http://news.nationalgeographic.com/news/2009/09/photogalleries/pacific-garbage-patch-pictures/

Hansen, Mark B. N. "System-Environment Hybrids." In Clarke and Hansen 2009, 113–142.

Hardin, Garrett. "Discriminating Altruisms." In Tobias 1984, 182–205.

Hardin, Garrett. *The Limits of Altruism: An Ecologist's View of Survival.* Bloomington: University of Indiana Press, 1977.

Harré, Rom. "The Siren Song of Substantivalism." *Journal for the Theory of Social Behaviour* 39, no. 4 (2009): 466–473.

Hastings, Alan, Carole L. Hom, Stephen Ellner, Peter Turchin, and H. Charles J. Godfrey. "Chaos in Ecology: Is Mother Nature a Strange Attractor?" *Annual Review of Ecology and Systematics* 24 (1993): 1–33.

Hayden, Patrick. "Gilles Deleuze and Naturalism: A Convergence with Ecological Theory and Politics." *Environmental Ethics* 19, no. 2 (Summer 1997): 185–204.

Hayles, Katherine. *My Mother Was a Computer: Digital Subjects and Literary Texts.* Chicago: University of Chicago Press, 2005.

Heidegger, Martin. *Basic Writings*. Edited by David Farrell Krell. Toronto: Harper Perennial, 2008.

Hendry, Andrew P., and Andrew Gonzalez. "Whither Adaptation?" *Biological Philosophy* 23 (2008): 673–699.

Herzog, Werner. *Encounters at the End of the World*, DVD. Directed by Werner Herzog. 2007. Montréal: Seville Pictures, 2007.

Herzog, Werner. *My Best Fiend*, DVD. Directed by Werner Herzog. 1999. Troy, Michigan: Anchor Bay Entertainment, 2002.

Herzogenrath, Bernd, ed. *Deleuze|Guattari and Ecology*. New York: Palgrave Macmillan, 2009.

Herzogenrath, Bernd. "The 'Weather of Music:' Sounding Nature in the Twentieth and Twenty-First Centuries." In Herzogenrath 2009, 216–232.

Hickman, Leo. "James Lovelock on the Value of Skeptics and Why Copenhagen Was Doomed." *The Guardian*. March 29, 2010. Accessed September 12, 2010. http://www.guardian.co.uk/environment/blog/2010/mar/29/james-lovelock.

Howe, Benjamin. "Was Arne Næss Recognized as the Founder of Deep Ecology Prematurely? Semantics and Environmental Philosophy." *Environmental Ethics* 32, no. 4 (Winter 2010): 369–385.

Hung, Ruyu. "Educating for and through Nature: A Merleau-Pontian Approach." *Studies in Philosophy and Education* 27 (2008): 355–367.

Husserl, Edmund. *Crisis of the European Sciences and Transcendental Phenomenology: An Introduction to Phenomenological Philosophy*. Translated by David Carr. Evanston: Northwestern University Press, 1970.

Husserl, Edmund. *Ideas Pertaining to a Pure Phenomenology and to a Phenomenological Philosophy I*. Translated by Fred Kersten. The Hague: Martinus Nijhoff, 1982.

Ingold, Tim. "Bindings and Boundaries: Entanglements of Life in an Open World." *Environment and Planning* 40 (2008): 1796–1810.

Ivanhoe, Philip J. "Of Geese and Eggs: In What Sense Should We Value Nature as a System?" *Environmental Ethics* 32, no. 1 (Spring 2010): 67–78.

Jantsch, Erich. *The Self-Organizing Universe: Scientific and Human Implications of the Emerging Paradigm of Evolution*. New York: Pergamon, 1980.

Jeffers, Robinson. *Not Man Apart: Photographs of the Big Sur Coast*. Edited by David Ross Brower. New York: Bristol Park Books, 1995.

Jelinski, Dennis E. "There Is No Mother Nature—There Is No Balance of Nature: Culture, Ecology, and Conservation." *Human Ecology* 33, no. 2 (April 2005): 271–288.

Jensen, Casper Bruun, and Kjetil Rödje, eds. *Deleuzian Intersections: Science, Technology, Anthropology*. New York: Berghahn Books, 2010.

Jickling, Bob, and Paul C. Paquet. "Wolf Stories: Reflections on Science, Ethics, and Epistemology." *Environmental Ethics* 27, no. 2 (Summer 2005): 115–134.

Jonas, Hans. *The Phenomenon of Life*. New York: Dell Publishing, 1966.

Joyce, James. *Finnegans Wake*. New York: Penguin Classics, 2000.

Kagawa-Fox, Midori. "Environmental Ethics from the Japanese Perspective." *Ethics, Place, and Environment* 13, no. 1 (2010): 57–73.

Kant, Immanuel. *Critique of the Power of Judgment*. Translated by Paul Guyer. New York: Cambridge University Press, 2000.

Kates, R. W., C. E. Colten, S. Laska, and S. P. Leatherman. "Reconstruction of New Orleans after Hurricane Katrina: A Research Perspective." *Cities and Rivers II: New Orleans, the Mississippi Delta, and Katrina* (August 29, 2006) http://www.ndsciencehumanitiespolicy.org/katrina/

Kaushik, Rajiv. "Physis and Flesh." In Tymieniecka 2009, 77–91.

Kazarian, Edward. "Deleuze, Perversion, and Politics." *International Studies in Philosophy* 30, no. 1 (1998): 91–106.

Kidner, David A. "Fraud, Fantasy, and Fiction in Environmental Writing." *Environmental Ethics* 27, no. 4 (Winter 2005): 391–410.

Kim Sung-Hae. "The Immoral World: The Telos of Daoist Environmental Ethics." *Environmental Ethics* 30, no. 2 (Summer 2008): 135–157.

Klenk, Nicole. "The Ethics of 'Following Nature' in Forestry: Academic Forest Scientists and Rolston's Environmental Ethics." *Environmental Ethics* 31, no. 1 (Spring 2009): 67–84.

Knyazeva, Helena. "Nonlinear Cobweb of Cognition." *Foundations of Science* 14, no. 3 (August 2009): 167–179.

Kohák, Erazim. *The Embers and the Stars: A Philosophical Inquiry into the Moral Sense of Nature*. Chicago: University of Chicago Press, 1984.

Konopka, Adam. "Ecological Goods that Obligate: A Husserlian Approach." *Environmental Ethics* 31, no. 3 (Fall 2009): 245–262.

Kowalski, Nathan. "Following Human Nature." *Environmental Ethics* 28, no. 2 (Summer 2006): 165–183.

Kretz, Lisa. "Open Continuity." *Ethics and the Environment* 14, no. 2 (Fall 2009): 115–137.

LaChappelle, Dolores. "Sacred Land, Sacred Sex." In Tobias 1984, 102–121.

Lamont, Michèle. *How Professors Think: Inside the Curious World of Academic Judgment*. Cambridge: Harvard University Press, 2009.

Latour, Bruno. *Science in Action: How to Follow Engineers and Scientists through Society*. Cambridge: Harvard University Press, 1987.

Latour, Bruno. *The Politics of Nature: How to Bring the Sciences into Democracy*. Translated by Catherine Porter. Cambridge: Harvard University Press, 2004.

Lawlor, Leonard. "The End of Phenomenology: Expressionism in Deleuze and Merleau-Ponty." *Continental Philosophy Review* 31 (1998): 15–34.

Lawlor, Leonard. *Thinking through French Philosophy: The Being of the Question*. Bloomington: Indiana University Press, 2003.

Lee So-Young. "Korean Environmental Thought and Practice: A Case Study of the Indramang Community." *Environmental Ethics* 30, no. 2 (Summer 2008): 115–134.

Leigh, Mike. *Happy-Go-Lucky*, DVD. Directed by Mike Leigh. 2011. Los Angeles: Miramax Lionsgate, 2008.

Leopold, Aldo. *A Sand County Almanac*. New York: Ballantine Books, 1970.

Levinas, Emmanuel. *Totality and Infinity: An Essay on Exteriority*. Translated by Alphonso Lingis. Pittsburgh: Duquesne University Press, 1969.

Lewontin, Richard. *The Triple Helix: Gene, Organism, and Environment*. Cambridge: Harvard University Press, 2000.

Light, Andrew. "Environmental Pragmatism as Philosophy or Metaphilosophy? On the Weston-Katz Debate." In *Environmental Pragmatism*. Edited by Andrew Light and Eric Katz, 325–334. New York: Routledge, 1996.

Llorente, Renzo. "The Moral Framework of Peter Singer's *Animal Liberation*: An Alternative to Utilitarianism." *Ethical Perspectives* 16, no. 1 (2009): 61–80.

Low, Douglas. "The Foundations of Merleau-Ponty's Ethics Theory." *Human Studies* 17, no. 2 (April 1994): 173–187.

Luhmann, Niklas. "Self-Organization and Autopoiesis." In Clarke and Hansen, 2009, 143–156.

Luhmann, Niklas. *Essays on Self-Reference*. New York: Columbia University Press, 1990.

Luhmann, Niklas. *The Reality of the Mass Media*. Translated by Kathleen Cross. Stanford: Stanford University Press, 2000.

Lulka, David. "Social Splinters and Cross-Cultural Leanings: A Cartographic Method for Examining Environmental Ethics." *Journal of Agricultural and Environmental Ethics* 21, no. 3 (2008): 275–296.

Madison, Gary Brent. *The Phenomenology of Merleau-Ponty: A Search for the Limits of Consciousness*. Athens: Ohio University Press, 1981.

Mann, Bonnie. "World Alienation in Feminist Thought: The Sublime Epistemology of Emphatic Anti-Essentialism." *Ethics and the Environment* 10, no. 2 (Fall 2005): 45–74.

Marangudakis, Manussos. "On Nature, Christianity, and Deep Ecology: A Response to W. S. Helton and N. D. Helton." *Journal of Moral Education* 37, no. 2 (2008): 245–248.

Marcelle, Daniel. "Making the Case for Gestalt Organization: Edmund Husserl and Aron Gurwitsch on the Problem of Independent Parts." In Nenon and Blosser 2010, 197–222.

Maskit, Jonathan. "Subjectivity, Desire, and the Problem of Consumption." In Herzogenrath 2009, 129–144.

Mathews, Freya. "Beyond Modernity and Tradition: A Third Way for Development." *Ethics and the Environment* 11, no. 2 (2006): 85–113.

Matthen, Mohan. "Chicken, Eggs, and Speciation." *Nous* 43, no. 1 (2009): 94–115.

Maturana, Humberto R. "The Nature of the Laws of Nature." *Systems Research and Behavioural Science* 17 (2000): 459–468.

Maturana, Humberto R., and Francisco Varela. *Autopoiesis and Cognition: The Realization of the Living*. Boston: D. Reidel, 1980.

Maturana, Humberto R., and Francisco Varela. *The Tree of Knowledge: The Biological Roots of Human Understanding*. Boston: Shambhala, 1987.

Max, D. T. "The Unfinished: David Foster Wallace's Struggle to Surpass Infinite Jest." *The New Yorker*. March 9, 2009. Accessed September 12, 2010. http://www.newyorker.com/reporting/2009/03/09/090309fa_fact_max

Mayeux, Raymond. "Mapping the New Frontier: Complex Genetic Disorders." *Journal of Clinical Investigation* 115, no. 6 (2005): 1404–7.

McGann, Marek. "Enactive Theorists Do It On Purpose: Toward an Enactive Account of Goals and Goal-Directedness." *Phenomenology and the Cognitive Sciences* 6, no. 4 (2007): 463–483.

McIntyre, Lee. "Making Philosophy Matter–or Else." *The Chronicle of Higher Education*. December 11, 2011. Accessed January 31, 2012. http://chronicle.com/article/Making-Philosophy-Matter-or/130029.

McLuhan, Marshall, and Eric McLuhan. *Laws of Media: The New Science*. Toronto: University of Toronto Press, 1988.

McShane, Katie. "Anthropocentrism vs Nonanthropocentrism: Why Should We Care?" *Environmental Values* 16, no. 2 (May 2007): 169–186.

Meillassoux, Quentin. *After Finitude: An Essay on the Necessity of Contingency*. New York: Continuum, 2008.

Merleau-Ponty, Maurice. *Nature: Course Notes from the Collège de France*. Translated by Robert Vallier. Evanston: Northwestern University Press, 2003.

Merleau-Ponty, Maurice. *The Phenomenology of Perception*. Translated by Colin Smith. New York: Routledge Press, 1962.

Merleau-Ponty, Maurice. *The Primacy of Perception*. Translated by James M. Edie. Evanston: Northwestern University Press, 1964.

Merleau-Ponty, Maurice. *The Visible and the Invisible*. Translated by Alphonso Lingis. Evanston: Northwestern University Press, 1968.

Michael, Mark A. "Is It Natural to Drive Species to Extinction?" *Ethics and the Environment* 10, no. 1 (Spring 2005), 49–66.

Minteer, Ben A. "Intrinsic Value for Pragmatists." *Environmental Ethics* 23, no. 1 (Spring 2001): 57–75.

Minteer, Ben A., and James P. Collins. "From Environmental to Ecological Ethics: Toward a Practical Ethics for Ecologists and Conservationists." *Science and Engineering Ethics* 14, no. 4 (December 2008): 483–501.

Minteer, Ben A., and James P. Collins. "Why We Need an Ecological Ethics." *Frontiers in Ecology and the Environment* 3, no. 6 (August 2005): 332–337.

Moore, Kathleen Dean. "The Truth of the Barnacles: Rachel Carson and the Moral Significance of Wonder." *Environmental Ethics* 27, no. 3 (Fall 2005): 265–277.

Morris, Simon Conway. *Life's Solution: Inevitable Humans in a Lonely Universe*. New York: Cambridge University Press, 2003.

Myers, Norman. "The Challenge of Disappearing Species." In Tobias 1984, 45–57.

Næss, Arne, and Ivar Mysterud. "Philosophy of Wolf Policies I: General Principles and Preliminary Exploration of Selected Norms." *Conservation Biology* 1, no. 1 (May 1987): 22–34.

Næss, Arne. "Heidegger, Postmodern Theory, and Deep Ecology." *Trumpeter* 14, no. 4 (1997): 2–7.

Næss, Arne. "Identification as a Source of Deep Ecological Attitudes." In Tobias 1984, 256–270.

Næss, Arne. "Living a Life that Reflects Evolutionary Insight." *Conservation Biology* 10, no. 6 (December 1996): 1557–1559.

Næss, Arne. "Reflections about Total Views." *Philosophy and Phenomenological Research* 25, no. 1 (September 1964): 16–29.

Næss, Arne. "Spinoza and Ecology." *Philosophia* 7, no. 1 (March 1977): 45–54.
Næss, Arne. "The Place of Joy in a World of Fact." *The North American Review* 258, no. 2 (Summer 1973): 53–57.
Næss, Arne. *Ecology, Community, and Lifestyle.* Translated by David Rothenberg. New York: Cambridge University Press, 1989.
Næss, Arne. *Freedom, Emotion, and Self-Subsistence: The Structure of a Central Part of Spinoza's Ethics.* Oslo: Universites for Laget, 1975.
Næss, Arne. *Gandhi and the Nuclear Age.* Totowa: Bedminster Press, 1965.
Næss, Arne. *The Call of the Mountain: Arne Næss and the Deep Ecology Movement.* Directed by Jan van Boeckel, and Pat van Boeckel. Blankenham, The Netherlands: ReRun Producties, 1997.
Næss, Arne. *The Selected Works of Arne Næss.* Edited by Alan Drengson. Dordrecht: Springer Netherlands, 2005.
Nagel, Thomas. "What Is It Like to Be a Bat?" In *Philosophy of Mind: A Guide and Anthology.* Edited by John Heil, 528–538. Oxford: Oxford University Press, 2004.
Naruse, Miyuko, and Takashi Iba. "Ecosystem as an Autopoietic System: Considering the Relationship between Ecology and Society Based on Luhmann's Theory." Paper presented at the Fourth Joint Japan-North America Mathematical Sociology Conference, Redondo Beach, California, May 31, 2008.
Nash, Roderick Frazier. *The Rights of Nature: A History of Environmental Ethics.* Madison: University of Wisconsin Press, 1989.
Nash, Roderick. "Rounding Out the American Revolution: Ethical Extension and the New Environmentalism." In Tobias 1984, 170–181.
Nenon, Thomas, and Phillip Blosser, eds. *Advancing Phenomenology: Essays in Honor of Lester Embree.* Dordrecht: Springer Netherlands, 2010.
Nichols, Shaun, Stephen Stich, and Jonathan M. Weinberg. "Normativity and Epistemic Intuitions." *Philosophical Topics* 11, no. 4 (December 2001): 429–460.
Nijenhuis, Ellert R. S., Johan Vanderlinden, and Philip Spinhoven. "Animal Defensive Reactions as a Model for Trauma-Induced Dissociative Reactions." *Journal of Traumatic Stress* 11, no. 2 (1998): 243–260.
Nishigaki, Toru. "The Ethics in Japanese Information Society: Consideration on Francisco Varela's *The Embodied Mind* from the Perspective of Fundamental Informatics." *Ethics and Information Technology* 8 (2006): 237–242.
Noë, Alva. *Action in Perception.* Cambridge: MIT Press, 2004.
Nolt, John. "The Move from *Is* to *Good* in Environmental Ethics." *Environmental Ethics* 31, no. 2 (Summer 2009): 135–154.
Norton, Bryan G. "Why I Am Not a Nonanthropocentrist: Callicott and the Failure of Monistic Inherentism." *Environmental Ethics* 17, no. 4 (Winter 2005): 341–358.
Nunez, Theodore W. "Rolston, Lonergan, and the Intrinsic Value of Nature." *The Journal of Religious Ethics* 27, no. 1 (Spring 1999): 105–128.
O'Hara, Ann M., and Fergus Shanahan. "The Gut Flora as a Forgotten Organ." *European Molecular Biology Organization Reports* 7, no. 7 (July 2006): 688–693.

O'Leary-Hawthorne, John. "The Bundle Theory of Substances and the identity of Indiscernibles." *Analysis* 55, no. 3 (July 1995): 191–196.

Orr, David W. *The Nature of Design: Ecology, Culture, and Human Intention*. Oxford: Oxford University Press, 2002.

Ott, Paul. "Value as Practice and the Practice of Value: Dewey's Value Theory for Environmental Ethics." *Environmental Ethics* 32, no. 3 (Fall 2010): 285–304.

Palmer, Clare. "Harm to Species? Species, Ethics, and Climate Change: The Case of the Polar Bear." *Notre Dame Journal of Law, Ethics, and Public Policy* 23, no. 2 (2009): 587–603.

Palmer, Craig. *When the Fish Are Gone: Ecological Disaster and the Fishers of Northwest Newfoundland*. Toronto: Fernwood Books, 2003.

Pang-White, Ann A. "Nature, Interthing Intersubjectivity, and the Environment: A Comparative Analysis of Kant and Daoism." *Dao: A Journal of Comparative Philosophy* 8, no. 1 (2009): 61–78.

Parisi, Luciana. "Technoecologies of Sensation." In Herzogenrath 2009, 182–199.

Parkes, Graham. "Nietzsche's Environmental Philosophy: A Trans-European Perspective." *Environmental Ethics* 27, no. 1 (Spring 2005): 77–91.

Parkinson, Alan, and Lynsie Chia Yin Chew, "The Use of Environmental Management Accounting Information by Managers in Companies in China: Theoretical Themes, Recognized Realities." Paper presented at the Sixth International Conference on Environmental, Cultural, Economic, and Social Sustainability, Cuenca, Ecuador, January 7, 2010.

Patten, Bernard C., and Gregor T. Auble. "A Systems Approach to the Concept of Niche." *Synthese* 43, no. 1 (January 1980): 155–182.

Peterson, Anna L. "Toward a Materialist Environmental Ethics." *Environmental Ethics* 28, no. 4 (Winter 2006): 375–393.

Peterson, M. Nils. "An Approach for Demonstrating the Social Legitimacy of Hunting." *Wildlife Society Bulletin* 32, no. 2 (Summer 2004): 310–321,

photo-news/photojournalism/e3i7c717df4a3ba58ebabe31ce5a5c47bbe.

Pickering, Andrew. "Cybernetics as a Nomad Science." In Jensen and Rödje 2010, 155–162.

Plumwood, Val. "Integrating Ethical Frameworks for Animals, Humans, and Nature: A Critical Feminist Eco-Socialist Analysis." *Ethics and the Environment* 5, no. 2 (2000): 285–322.

Plumwood, Val. "The Concept of a Cultural Landscape: Nature, Culture, and Agency in the Land." *Ethics and the Environment* 11, no. 2 (2006): 115–150.

Plumwood, Val. *Environmental Culture: The Ecological Crisis of Reason*. New York: Routledge, 2002.

Plumwood, Val. *Feminism and the Mastery of Nature*. New York: Routledge, 1993.

Precoda, Norman. "Requiem for the Aral Sea." *Ambio* 3, no. 4 (May 1991): 109–114.

Preston, Christopher J., and Steven H. Corey. "Public Health and Environmentalism: Adding Garbage to the History of Environmental Ethics." *Environmental Ethics* 27, no. 1 (Spring 2005): 3–22.

Prinz, Jesse, and Andy Clark. "Putting Concepts to Work: Some Thoughts for the Twentyfirst Century." *Mind and Language* 19, no. 1 (February 2004): 57–69.

Protevi, John. "Beyond Autopoiesis: Inflections of Emergence and Politics in Francisco Varela." In Clarke and Hansen 2009, 94–112.

Purpura Jr., Gary J. "In Search of Human Uniqueness." *Philosophical Psychology* 19, no. 4 (2006): 443–461.

Putnam, Hilary. "Meaning and Reference." *The Journal of Philosophy* 70, no. 19 (November 1973): 699–711.

Quilley, Stephen. "The Land Ethic as an Ecological Civilizing Process: Aldo Leopold, Norbert Elisa, and Environmental Philosophy." *Environmental Ethics* 31, no. 2 (Summer 2009): 115–134.

Rancière, Jacques. *On the Shores of Politics*. Translated by Liz Heron. New York: Verso, 2007.

Reeve, Jennifer. "Long-Term Sustainability of a Low-Input Organic Farm." Paper presented at the Sixth International Conference on Environmental, Cultural, Economic, and Social Sustainability, Cuenca, Ecuador, January 5, 2010.

Riggio, Adam. "John Dewey as a Philosopher of Contingency and the Value of This Idea for Environmental Philosophy." *Environmental Ethics* 33, no. 4 (Winter 2012): 395–413.

Riis, Søren. "Towards the Origin of Modern Technology: Reconfiguring Martin Heidegger's Thinking." *Continental Philosophy Review* 44 (2011): 103–117.

Roach, Catherine M., Tim I. Hollis, Brian E. McLaren, and Dean L. Y. Bavington. "Ducks, Bogs, and Guns: A Case Study of Stewardship Ethics in Newfoundland." *Ethics and the Environment* 11, no. 1 (Spring 2006): 43–70.

Robert, Jason Scott. "Wild Ontology: Elaborating Environmental Pragmatism." *Ethics and the Environment* 5, no. 2 (2000): 191–209.

Rogers, Raymond A. "Are Environmentalists Hysterical or Paranoid? Metaphors of Care and 'Environmental Security.'" *Ethics and the Environment* 5, no. 2 (2000): 211–227.

Rogers, Tim B. "Nature of the Third Kind: Toward an Explicitly Relational Constructivism." *Environmental Ethics* 31, no. 4 (Winter 2009) 393–412.

Rokstad, Konrad. "Was Merleau-Ponty a Phenomenologist? Some Reflections upon the Identity of Phenomenology." In Tymieniecka 2009, 3–36.

Rolston, Holmes III. "A Managed Earth and the End of Nature." In *Research in Philosophy and Technology: Philosophies of the Environment and Technology*, edited by Marina Paolo Banchetti-Robino, Don E. Marietta Jr., and Lester Embree, 143–164. Stanford: JAI Press, 1999.

Rolston, Holmes III. "Caring for Nature: From Fact to Value, from Respect to Reverence." *Zygon* 39, no. 2 (June 2004): 277–302.

Rolston, Holmes III. "F/Actual Knowing: Putting Facts and Values in Place." *Ethics and the Environment* 10, no. 2 (2005): 137–174.

Rolston, Holmes III. "Mountain Majesties above Fruited Plains: Culture, Nature, and Rocky Mountain Aesthetics." *Environmental Ethics* 30, no. 1 (Spring 2008): 3–20.

Rolston, Holmes III. *Conserving Natural Value.* New York: Columbia University Press, 1994.

Rorty, Richard. *Contingency, Irony, and Solidarity.* New York: Cambridge University Press, 1989.

Rothenberg, David. *Is It Painful to Think? Conversations with Arne Næss.* Minneapolis: University of Minnesota Press, 1993.

Sancton, Julian. "Mankind Creates First Synthetic Genome, Officially Replaces God." *Vanity Fair.* May 21, 2010. Accessed July 12, 2010. http://www.vanityfair.com/online/daily/2010/05/mankind-creates-first-synthetic-genome-officially-replaces-god.html.

Sarris, Viktor. "Relational Psychophysics: Messages from Ebbinghaus' and Wertheimer's Work." *Philosophical Psychology* 23, no. 2 (2010): 207–216.

Sartre, Jean-Paul. *Existentialism Is a Humanism.* Translated by Carol Macomber. New Haven: Yale University Press, 2007.

Sartre, Jean-Paul. *Nausea.* Translated by Lloyd Alexander. New York: New Directions Publishing, 1964.

Schaffer, William M. "Order and Chaos in Ecological Systems." *Ecology* 66, no. 1 (February 1985): 93–106.

Schaffer, William M., and M. Kot. "Do Strange Attractors Govern Ecological Systems?" *BioScience* 35, no. 6 (June 1985): 342–350.

Schrader-Frechette, Kristin. "Individualism, Holism, and Environmental Ethics." *Ethics and the Environment* 1, no. 1 (1996): 55–69.

Schrift, Alan D. "Deleuze Becoming Nietzsche Becoming Spinoza Becoming Deleuze: Toward a Politics of Immanence." *Philosophy Today* 50 (2006): 187–194.

Searle, John R. *Mind: A Brief Introduction.* Oxford: Oxford University Press, 2004.

Sehdev, Mina. "Perception, Textual Theory, and Metaphorical Language." In Tymieniecka, 2009, 233–240.

Semetsky, Inna. "The Magician in the World: Becoming, Creativity, and Transversal Communication." *Zygon* 44, no. 2 (2009): 323–345.

Serres, Michel. *The Natural Contract.* Translated by Elizabeth MacArthur and William Paulson. Ann Arbor: University of Michigan Press, 1995.

Sessions, George. "Ecological Consciousness and Paradigm Change." In Tobias 1984, 28–44.

Shepard, Paul. "If You Care about Nature You Can't Go on Hating the Germans Like This." In Tobias 1984, 206–212.

Siebert, Charles. "The Animal Cruelty Syndrome." *New York Times*, June 7, 2010, Magazine Section, International Edition.

Siipi, Helena. "Dimensions of Naturalness." *Ethics and the Environment* 13, no. 1 (2008): 71–103.

Simus, Jason. "Environmental Art and Ecological Citizenship." *Environmental Ethics* 30, no. 1 (Spring 2008): 21–36.

Singer, Peter. *The Expanding Circle: Ethics and Sociobiology.* New York: Farrar, Straus, and Giroux, 1981.

Skakoon, Elizabeth. "Nature and Human Identity." *Environmental Ethics* 30, no. 1 (Spring 2008): 37–49.

Smith, Mick. *An Ethics of Place: Radical Ecology, Postmodernity, and Social Theory.* Albany: State University of New York Press, 2001.

Smolin, Lee. *The Life of the Cosmos.* New York: Oxford University Press, 1997.

Soulé, Michael E. "What Is Conservation Biology?" *Bioscience,* 35, no. 11 (December 1985): 727–734.

Sowards, Stacey K. "Identification through Orangutans: Destabilizing the Nature/Culture Dualism." *Ethics and the Environment* 11, no. 2 (2006): 45–61.

Spinoza, Benedict. *The Ethics, Treatise on the Emandation of the Intellect, Selected Letters.* Edited by Seymour Feldman. Translated by Samuel Shirley. Indianapolis: Hackett, 1992.

Starkey, Charles. "The Land Ethic, Moral Development, and Ecological Rationality." *Southern Journal of Philosophy* 45, no. 1 (Spring 2007): 149–175.

Stengers, Isabelle. "Experimenting with *What Is Philosophy?*" In Jensen and Rödje 2010, 39–56.

Sterba, James P. "Biocentrism and Human Health." *Ethics and the Environment* 5, no. 2 (2000): 271–284.

Stevens, Piers H. G. "Toward a Jamesian Environmental Philosophy." *Environmental Ethics* 31, no. 3 (Fall 2009): 227–244.

Stockton, Will. "Reading Like a Sodomite: Deleuze, Donne, Eliot, Presentism, and the Modern Renaissance." *Rhizomes: Cultural Studies in Emerging Knowledge* 17 (Winter 2008). Accessed June 26, 2011. http://www.rhizomes.net/issue17/stockton.html.

Stone, Christopher D. "Should Trees Have Standing? Toward Legal Rights for Natural Objects." *Southern California Law Review* 45 (Spring 1972): 450–501.

Stroll, Avrum. *Twentieth Century Analytic Philosophy.* New York: Columbia University Press, 2000.

Sullins, John. "Transcending the Meat: Immersive Technologies and Computer Mediated Bodies." *Journal of Experimental and Theoretical Artificial Intelligence* 12 (2000): 13–22.

Sullivan, Shannon. *Living Across and through Skins: Transactional Bodies, Pragmatism, and Feminism.* Bloomington: Indiana University Press, 2001.

Svanæs, Dag. "Context-Aware Technology: A Phenomenological Perspective." *Human-Computer Interaction* 16 (2001): 379–400.

Swanton, Christine. *Virtue Ethics: A Pluralist View.* Oxford: Oxford University Press, 2003.

Tanner, Julia K. "The Argument from Marginal Cases and the Slippery Slope Objection." *Environmental Values* 18, no. 1 (February 2009): 51–66.

Taylor, Paul W. *Respect for Nature: A Theory of Environmental Ethics.* Princeton: Princeton University Press, 1986.

Thompson, Evan. "Life and Mind: From Autopoiesis to Neurophenomenology, a Tribute to Francisco Varela." *Phenomenology and the Cognitive Sciences* 3 (2004): 381–398.

Thompson, Evan. *Mind in Life: Biology, Phenomenology, and the Sciences of the Mind.* Cambridge: Harvard University Press, 2007.

Tierney, John. "Do You Have Free Will? Yes, It's the Only Choice." *New York Times,* March 22, 2011. Accessed March 23, 2011. http://www.nytimes.com/2011/03/22/science/22tier.html.

Toadvine, Ted. "Ecophenomenology and the Resistance of Nature." In Nenon and Blosser 2010, 343–355.

Toadvine, Ted. "Limits of the Flesh: The Role of Reflection in David Abram's Ecophenomenology." *Environmental Ethics* 27, no. 2 (Summer 2005): 155–170.

Tobias, Michael, ed. *Deep Ecology.* San Marcos: Avant Books, 1984.

Tobias, Michael. "Humanity and Radical Will: Reflections from the Island of Life." In Tobias 1984, 2–27.

Todres, Leslie A. "Embracing Ambiguity: Transpersonal Development and the Phenomenological Tradition." *Journal of Religion and Health* 39, no. 3 (Fall 2000): 227–237.

Todt, Oliver, Javier Rodríguez Alcázar, and José Luis Luján. "Practical Values and Uncertainty in Regulatory Decision-Making." *Social Epistemology* 24, no. 4 (2010): 349–362.

Tymieniecka, A. T., ed. *Phenomenology and Existentialism in the Twentieth Century: Fruition, Cross-Pollination, Dissemination.* Dordrecht: Springer Netherlands, 2009.

Valsiner, Jaan, and René van der Veer. *The Social Mind: Construction of the Idea.* Cambridge: Cambridge University Press, 2000.

Varela, Francisco J. "The Early Days of Autopoiesis." In Clarke and Hansen 2009, 62–76.

Varela, Francisco J., Evan Thompson, and Eleanor Rosch. *The Embodied Mind: Cognitive Science and Human Experience.* Cambridge: MIT Press, 1991.

Varner, Gary. "Utilitarianism and the Evolution of Ecological Ethics." *Science and Engineering Ethics* 14, no. 4 (2008): 551–573.

Velasco, Joel D. "When Monophyly Is Not Enough: Exclusivity as the Key to Defining a Phylogenetic Species Concept." *Biology and Philosophy* 24, no. 4 (September 2009): 473–486.

Verley, Xavier. "Cosmologie et Phénoménologie: Whitehead et Merleau-Ponty." *Revue Philosophique de la Franca et L'Étranger* 131, no. 1 (2006): 35–55.

Volosinov, V. N. *Marxism and the Philosophy of Language.* New York: Seminar Press, 1973.

von Uexküll, Jakob. "The Theory of Meaning." *Semiotica* 41, no. 1 (1982): 25–82.

von Uexküll, Jakob. *Striefzüge durch die Umwelten von Tieren und Menschen.* Hamburg: Rowohlt Taschenbuch Verlag, 1956.

Wade, Nicholas. "Researchers Say They Created a 'Synthetic Cell.'" *New York Times.* May 21, 2010. Accessed July 12, 2010. http://www.nytimes.com/2010/05/21/science/21cell.html.

Walker, David. "Lu Guang Wins National Geographic Photography Grant." *Photo District News,* May 27, 2010. Accessed September 5, 2010. http://www.pdnonline.com/pdn/content_display/

Walker, Margaret Urban. *Moral Understandings: A Feminist Study in Ethics*. New York: Routledge, 1998.
Ward, Peter. *The Medea Hypothesis: Is Life on Earth Ultimately Self-Destructive?* Princeton: Princeton University Press, 2009.
Weber, Andreas, and Francisco Varela. "Life after Kant: Natural Purposes and the Autopoietic Foundations of Biological Individuality." *Phenomenology and the Cognitive Sciences* 1 (2002): 97–125.
Weber, Thomas. "Gandhi, Deep Ecology, Peace Research, and Buddhist Economics." *Journal of Peace Research* 36, no. 3 (May 1999): 349–361.
Weil, Simone. *The Need for Roots: Prelude to a Declaration of Duties towards Mankind*. London: Routledge, 1952.
Weinberg, Jonathan M. "How to Challenge Intuitions Empirically without Risking Skepticism." *Midwest Studies in Philosophy* 31 (2007): 318–343.
Weisman, Alan. *The World without Us*. Toronto: HarperCollins, 2008.
Williams, Malcolm. "Contingent Realism–Abandoning Necessity." *Social Epistemology* 25, no. 1 (2011): 37–56.
Williamson, Timothy. *The Philosophy of Philosophy*. Malden, MA: Blackwell Publishing, 2007.
Wolfe, Carey. "Meaning as Event-Machine, or Systems Theory and the 'Reconstruction of Deconstruction:' Derrida and Luhmann." In Clarke and Hansen 2009, 220–245.
Wolfe, Carey. *What Is Posthumanism?* Minneapolis: University of Minnesota Press, 2010.
Wringe, Bill. "Global Obligations and the Agency Objection." *Ratio: An International Journal of Analytic Philosophy* 23, no. 2 (June 2010): 217–231.
Zepke, Stephen. "Eco-Aesthetics: Beyond Structure in the Work of Robert Smithson, Gilles Deleuze, and Félix Guattari." In Herzogenrath 2009, 200–215.
Zimmerman, Michael E. "The Threat of Ecofascism." *Social Theory and Practice* 21, no. 2 (Summer 1995): 207–238.
Zimmerman, Michael E. *Heidegger's Confrontation with Modernity: Technology, Politics, and Art*. Bloomington: Indiana University Press, 1990.
Zizek, Slavoj. "Unbehagen in der Natur: Ecology against Nature." *Bedeutung Magazine* 1 (2008): 42–57.

INDEX

The manufacturer's authorised representative in the EU is Springer Nature Customer Service Centre GmbH, Europaplatz 3, 69115 Heidelberg, Germany. If you have any concerns regarding our products, please contact ProductSafety@springernature.com

Printed and bound by CPI Group (UK) Ltd, Croydon, CR0 4YY
13/10/2026
02232375-0017